THE NEW FARMER'S ALMANAC

2017

[THE COMMONS]

greenhorns

THE NEW FARMER'S ALMANAC 2017 [THE COMMONS]

Published by the Greenhorns
www.greenhorns.net

creative commons

Non-commercial attribution
(CC by NC)
All images are in the public domain unless noted
in the image credits.
To reprint original work, request permission of
the contributor.

Printed in the USA by Versa Press

ISBN 978-0-9863205-1-4

EDITOR IN CHIEF
Severine von Tscharner Fleming

LEAD EDITOR
Nina Pick

VISUAL EDITOR AND DESIGN
Katie Eberle

VISUAL EDITOR
Rodolph de Salis

LAYOUT
Nina Pick and Katie Eberle

COPY EDITOR AND PROOFREADER
Christopher Church

FRONT COVER ILLUSTRATION
Amy Franceschini

INTERIOR COVER ILLUSTRATION
Allison Howe

CHAPTER ILLUSTRATIONS
Rachel Alexandrou

THANK YOU to all who wrote essays and sent illustrations and history snippets and poems. Thank you all who enjoy the book and took the time to praise us. Thank you to our sponsor, Organic Valley. Thank you Flora Family Foundation. Thank you to TomKat Ranch Foundation and Hope Foundation for your support to cover editorial, layout, copyediting, printing, and shipping costs; we really hope you are pleased and will support it again next year!

Thank you to Charlie Macquarie for your amazing leadership and Nina Pick for muscling in there. Thank you to *The Idler, Slightly Foxed, Inverness Almanac, The Land Magazine, Fedco Catalog, Common Ground, Dark Mountain Project, Farming Magazine*, and others for your publication inspiration. Thank you to Stephanie Mills for the encouragement and archival illuminations. Thank you to Tom Giessel, historian of the National Farmers Union, for once again trawling the archives. Thank you to Rick and Megan Prelinger for creating the context in which this book is possible—for assembling the archive, creating the habitat, and sugaring us all with inspiration.

Thank you to my mother, Renata von Tscharner, for use of the stockli to write the introduction, and to Davis Pike for housing me in Maine during the final days. Thank you to the Greenhorns team—especially Laura del Campo and Inés Chapela who have finessed the leadership transition I so badly needed. Thank you to the Peters family of Santa Fe, for hosting me at the little plum ranch in Rio en Medio. To Barbara and Barry Deutsch for their indefatigable championship of all Greenhorns happenings, for their hospitality, nurture, book suggestions and whole-ecosystem approach to everything.

Thank you to Ignacio Chapela who first counseled me to work with artists, and to Amy Franceschini who has gracefully and briskly designed all our websites and most of our logos for Greenhorns, Agrarian Trust, National Young Farmers Coalition, OUR LAND.tv, and Farm Hack, and whose new Future Farmers Project, Seed Journey, provides such an inspiring platform for my 2017 hiatus "at large" in the world. Long live Greenhorns! —SvTF

Many thanks to the whole Greenhorns team, and especially Laura del Campo, Bonnie Rubrecht, and Inés Chapela; Mesa Refuge, which has nurtured, on several occasions, the creative spirits of our various editors; Charlie Macquarie, Lead Editor Emeritus, for gracefully passing the baton and sharing so generously his knowledge and experience; Severine, for the inspiring, wide-open lens of her vision; the *Inverness Almanac* crew, for the love. —NP

TABLE OF CONTENTS

OCTOBER: LAND AND OWNERSHIP

NOVEMBER: GENETICS

DECEMBER: INTERGENERATIONAL DYNAMICS AND HISTORY

ABOUT THE GREENHORNS 352

INDEX OF CONTRIBUTORS

JASON DETZEL
Grass farmer, extension educator
Hudson, NY
facebook.com/DiamondHillsFarm

BERNADETTE DIPIETRO
Visual artist, writer
Ojai, CA
bernadettedipietro.com

THOMAS DRISCOLL
Director of Conservation Policy and Education
National Farmers Union
Washington, DC
nfu.org

LUCAS FARRELL
Goat farmer, writer
Townshend, VT
bigpicturefarm.com

HARRIET FASENFEST
Teacher, author, urban farm mom
Portland, OR

MICHAEL FOLEY
Farmer, writer, activist
Willits, CA

AMY FRANCESCHINI
Artist
San Francisco, CA
futurefarmers.com

GRAISON S. GILL
Baker, miller
New Orleans, LA
bellegardebakery.com

MARÍA JOSÉ GIMÉNEZ
Translator, poet
Easthampton, MA and Montreal
mariajosetranslates.com

SARAH GITTINS
Artist
Dundee and Fife, Scotland
sarahgittins.wordpress.com

TRISTAN GOOLEY
Author, natural navigator
West Sussex, UK
www.naturalnavigator.com

JILL HAMMER
Rabbi, writer, scholar, priestess
New York, NY
rabbijillhammer.com
kohenet.org

JEREMY HARRIS
Musician, recordist
Inverness, CA
jeremyrobertharris.com

LORIG HAWKINS
Farmer, writer
Austin, TX
middleground-farm.com

NICK HAYES
Graphic novelist, illustrator, political cartoonist
London, UK
foghornhayes.com

CHRISTINE HEINRICHS
Author
Cambria, CA
christineheinrichs.com

ELIZABETH HENDERSON
Farmer, writer
Newark, NY
thepryingmantis.wordpress.com

PARKER HIGGINS
Activist, writer
San Francisco, CA
parkerhiggins.net

THE HOP PROJECT
General Public (artists Chris Poolman
and Elizabeth Rowe)
Birmingham, UK
generalpublic.org.uk

SUSANNAH BRUCE HORNSBY
Writer, beekeeper
Yancey Mills, VA
featherbyfeather.com

ALLISON HOWE
Flower farmer, artist
Warner, NH

JOHN IKERD
Professor emeritus, Agricultural Economics
University of Missouri
Fairfield, IA
johnikerd.com

JOHN JEAVONS
Environmentalist, farmer, nonprofit director, author
Willits, CA
growbiointensive.org
johnjeavons.info

DERRICK JENSEN
Writer, activist
Crescent City, CA
derrickjensen.org

GRANT RICHARD JONES
Farmerbard, landscape architect, poet
Ellisforde, WA
jonesandjones.com

BINYAMIN KLEMPNER
Writer, farmer, activist
Jerusalem, Israel
growinggreens.net

ELIZABETH KOHLER
LMT, farmhand, writer
Bar Harbor, ME

BENNETT KONESNI
Farmer-musician
Belfast, ME
duckbackfarm.com
worksongs.org

ALI KRUGER
Community builder-at-large
Ann Arbor, MI

BRIAN LAIDLAW
Troubadour
Boulder, CO
brianlaidlaw.com

JOHN FRANCIS MCGILL
Poet and educator
North Bennington, VT

JANE MEAD
Grape grower, teacher
Napa County, CA, and at large
janemead.com

NEW MEXICO ACEQUIA ASSOCIATION
Santa Fe, NM
lasacequias.org

DANJO PALUSKA
Postcard aficionado
Brunswick, ME

BARBARA PLEASANT
Garden writer
Floyd, VA
barbarapleasant.com

MARTIN POWELL
Poet, activist
London, UK
martinpowellpoetry.com

RYAN POWER
Farmer
Sebastopol, CA
thenewfamilyfarm.com

QUEEN MOB COLLECTIVE
Pirates
#mpbsquatting

ERICA ROMKEMA
Community farm manager
Efland, NC

ABBY SADAUCKAS
Farmer
Bowdoinham, ME
applecreekfarm.wordpress.com

MARTHA SHAW
Artist
Great Barrington, MA

LAINE V. SHIPLEY
Herbalist, artist, green witch
Portland, OR

BEN SHORT
Charcoal burner, woodman, writer
Eggardon Hill, Dorset, England

LEAH SIENKOWSKI
Writer, fieldhand
Ada, MI
dreamgoats.com

ALLYSON SIWIK
Conservationist
Silver City, NM
gilaconservation.org

BREN SMITH
Ocean farmer
Thimble Islands, CT
greenwave.org

DYLAN SMITH
Scientist
Oakland, CA

JOHN ERIC SNIDER
Marketing agronomist, PGGSeeds
pggseeds.us

THOMAS SPENCE
Revolutionary (1750–1814)
Newcastle, England

STEVE SPRINKEL
Certified Organic Farmer
Ojai, CA
farmerandcook.com

SPURSE
Ecosystem Design and Consultation Collective
USA
spurse.org

CHARLOTTE XOCHITL CAMMER SULLIVAN
Citizen designer, expedition artist
North Bennington, VT
charlottexcsullivan.com

T. CODY SWIFT, MA
Counselor, qualitative researcher, and board member
The Heffter Research Institute and Riverstyx Foundation
Santa Cruz, CA
heffter.org
riverstyxfoundation.org

OLIVIA S. TINCANI
Food and farm business educator
Nomadic: California, Italy, and everywhere in between
oliviatincaniandco.com

DANIEL TUCKER
Islander
San Juan Island, WA
dantucker.wordpress.com

BEATRICE VERMEIR
Wordworker, apronmaker
London, UK
csm-arts.academia.edu/beatricevermeir

HOWARD WATTS III
Communications specialist,
Great Basin Water Network
Las Vegas, NV
greatbasinwater.net

BONNIE ROSE WEAVER
Urban farmer, community herbalist
San Francisco, CA
1849medicinegarden.com

RACHEL NICOLE WEAVER
Philosopher, agrarian
Denton, TX

CORY WILLIAM WHITNEY
Human ecologist
Center for Development Research (ZEF)
University of Bonn
World Agroforestry Center (ICRAF), Nairobi

ALMANACK
ALMANACK
ALMANACK
ALMANACK
ALMANACK
ALMANACK
ALMANACK

✻ ◆ ✳) ◉ ❁ ✦ ⦿ ⦂ ✴ ⊙ ⦾ ✳ ✻ ❊ ❀ ♈ ♔ ▫ ✿ ♉ ❦ ♊ ♔ ✾ (⧳ ⎰ (⎱ ? ♍)

EDITOR'S NOTE

NINA PICK, LEAD EDITOR

*Multitude stands in my mind but I think that the ocean in the bone vault is only
The bone vault's ocean: out there is the ocean's;
The water is the water, the cliff is the rock, come shocks and flashes of reality. The mind
Passes, the eye closes, the spirit is a passage;
The beauty of things was born before eyes and sufficient to itself; the heartbreaking beauty
Will remain when there is no heart to break for it.*
—Robinson Jeffers

The Old Farmer's Almanac was first printed in 1792 and has been published continuously ever since. The first Greenhorns' *New Farmer's Almanac* was published more than two hundred years later, in 2013. When I tell people about *The New Farmer's Almanac*, I often receive the response, "Oh cool, my grandfather used to read that." "Well, probably not," I say, "but yeah, kind of." And it's true. *The New Farmer's Almanac* is the new-old, and the old-new. Part of a cultural turn toward our human heritage, like learning Yiddish, or cooking with fire.

An almanac is an homage to time. The earliest usage of the word *almanac* was connected to astronomical calendars, and the almanac form continues to present a way of time-tending that is attuned to nonlinear temporality, to the cyclic and spiraling movement of stars, days, weather. An almanac marks planting and harvesting, birth and death, and all the living and working and eating and loving that goes on in between. Its traditional offering of recipes, calendars, cleaning tips, seed catalogs, snow predictions, and moon cycles restores us to an awareness of our place in nature, in our bodies, and in our homes and communities. For all of its emphasis on history and tradition, an almanac serves to remind us of our place in present time.

▫ ❦ ✳ ⚘ ✺ ♔ ❁ ❧ ✳ ❀ ✳) (✦ ✾ ❦ ❀ ▫ ✻ ♉ ❁ ✾ ⚘ ❦ ✻ ❁ ✾ ♔ ❁ ✦

The poet Joseph Brodsky wrote, "I have always adhered to the idea that God is time." In the divinity of time we see, not the fetishization that is the trademark of capitalism—the endless accumulation, acceleration, and expenditure—but time as presence. Time for Time's sake. The uncomfortable and beautiful realness of being for a moment with things as they are. The real that comes with facing into discomfort, nuance, and ambivalence. The real of cultivating a garden, of slaughtering an animal. The real of wet soil and dry sand. Stone, water, ice, wind. The smell of fire, of spring. God is one way to name this confrontation with reality.

In a cultural landscape characterized by screens, and the accompanying digitization of everything from sex and dating to language and landscape, farming is a way of practicing a commitment to the real. It is a practice like meditation is a practice. Like being kind and showing up for others is a practice. Engaging with the food system on the most basic and elemental level recalls us to our essential mode of being—it literally grounds us, and keeps our hands in the dirt.

It may be that farmers, and poets, and farmer-poets, of which there are many represented in this volume, are doing the most important work of our era, in both the visible and unseen realms. By tending to the soil and the soul (and these are not separate), they hold an essential cultural and human archive, a depth of meaning, and a mode of being that is essential to our society and is in danger of rapidly becoming lost.

As the philosopher Albert Borgmann stated, "Today the critical and crucial distinction for nature and humans in not between the real and the artificial but between the real and the hyperreal." The flickering back and forth between the real and the hyperreal—perhaps the word should be *hyporeal*—is a defining feature of our cultural moment. The commitment to being in this struggle, to continually assert our humanness in the face of digital technologies, is an exercise in freedom and dignity—the dignity to choose the object of our attention, rather than becoming lost in the delusion that characterizes addiction. This is a ongoing opportunity to opt for reality over the convenient, the prepackaged, and the comfortable. To choose real, messy, beautiful, loving, heartbreaking relationships over the ease of clicking "Like." Real meat with its visible blood and bones instead of the hygienic suffering available shrinkwrapped at the grocery store; real lettuce with dirt at the base of its leaves. Cold, wet sand on a foggy beach instead of a year-round, housebound 68 degrees. Real night, and real stars, instead of the mindless habit of porch lights. Glass instead of plastic. No packaging instead of plastic. Real clouds. Real twitter. Real kindling and tinder and the Amazon River. (May we tend to language as we tend to the land, and be on guard against those who would take the language of nature and modify it genetically and sell it back to us as a product for which we have no need but are convinced we cannot live without).

My work has centered around almanacs

because I believe, in essence, that an almanac is a love letter to the real. It is a love letter to everything of nature that is unpredictable and ungovernable and uncomfortable: weather; relationship; the mysterious drives within a seed that make it sprout according to its own schedule; the force of a netted fish struggling in the most important moment of its singular existence; the breath that pulses out of an animal the moment it transforms from living flesh into dinner in your hands. An almanac is a testimonial to intricacy, to interdependency, to intractable and unanswerable conundrums. It is a love letter to being human. It is a love letter to intimacy itself.

The Italian poet Cesare Pavese wrote, "I am not in league with inventors or adventurers, nor with travelers to exotic destinations. The surest—also the quickest—way to awake the sense of wonder in ourselves is to look intently, undeterred, at a single object. Suddenly, miraculously, it will reveal itself as something we have never seen before." Reality: the closer you get, the more magical it becomes. The form of the almanac—the hodgepodge, the stew made from whatever is left in the fridge, the everything-but-the-kitchen-sink aesthetic—is a mirror of its content, and is meant exactly to awaken this sense of wonder. It tells us, Everything's connected. Profound or mundane, nothing falls outside the map of the world. Which is to say: Living creature, you too are a part of it. Belong.

THE CEREALS OF THE LAKE DWELLINGS

INTRODUCTION

SEVERINE VON TSCHARNER FLEMING, EDITOR IN CHIEF

WELCOME TO THE ALMANAC

This *Almanac* follows the commons, not with linear arguments, but rather in an uncoiling elliptical orbit around the subjects dear to our hearts. A loose set of overlapping coils—what Hispano-Westerners might call a lariat. For working agrarians whose literary outlets become constrained by physical tiredness—our format is for you. We specialize in two to three pages for bedtime or outhouse reading. Take it as it comes.

Derrick Jensen, deep ecologist, urges his readers to destroy linear thought as a product of empire, and instead to focus on stories of relation and reciprocity, in open wonderment of interspecies happenings conducted in the everyday. You will find the themes of this *Almanac* follow this guidance.

Join us as we cherish the present tense as a form of rebellion. Jensen encourages us also to remove the dams that block the salmon, so that life can circle and cycle according to its higher genius flux, and itself author spontaneous revisions. To be ready for spontaneous revisions in our own lives and lifeways, to figure out the infrastructures of a different pattern of relation. This kind of flexibility, seems to suit the punk temperament of land workers, particularly

during an era of seismic climatic-political-eco-social flux, the highly potent now.

Welcome to our open miscellany of rambling, lyrical, opinionated, and unacademic riffs on land theory and political economy; welcome to a format with room for many opinions. The *Almanac* gives us a place to practice our rhetoric, our research practices, our personal and collective oratory. This *Almanac* is a modest stage for dancing out the notions we mull over in the sunshine, read about in dusty books, and see for ourselves in the lay of the land.

THE GREENHORNS

The *Almanac* is a large undertaking for the small production house called the Greenhorns, a nine-year-old grassroots organization based in the Champlain Valley of New York. The Greenhorns celebrates agrarian futurism; we work to promote, support, and recruit the hearts, brains, bodies, and businesses that it takes to build a more autonomous, diverse, prosperous farm economy. We aim for a world where regions and watersheds can eat their own food and drink their own water. We know this will take team work and will power. The Greenhorns celebrates the literacy, sensitivity, and intergenerational collaborations of a place-based culture and hopes to create rich contexts for this celebration. To this end, the Greenhorns produce many kinds of media and transmedia, from sailboat trade stunts to sing-along grange tours.

If you like this book, you are likely to enjoy our eight years of radio podcasts, our lively blog, our digital resource maps, our documentary film, our short Web film series, our cooperative film festival, our anthology, our various mix tapes and oral history projects, our vinyl record of grange songs, our many guidebooks and posters and screen-printed bling packs on Etsy. These were all created in collaboration. Your ideas could be next.

PREDICAMENT

What a twisted political season of violence and incongruence, endless interruptions and an uncanny sensation of imminent tyranny. Surveillance, drones, mergers, superpowers concentrating, financiers with metadata speculating. Big unresolved questions, unbearably high stakes, urgency, stasis, hype, dysfunction, but no clarity, no transparency; few are taking responsibility. Jill Stein is a fun exception.

How will it play out? Will the overflooding of pipelines, borders, rivers, cities stop anytime soon? What about the rise of the gun against innocents, children, minorities? Assaults against bodies and water bodies? Cancers terrorizing families? With each record storm, the sound of chainsaws, cutting of pavement, the sounds of roads, drains, culverts being repaired, enlarged. The deforestation of highway margins to make room for the predicted record snowfall, bankrupted municipalities, hate speech and traffic.

Fluctuating and unpredictable crisis. What can we believe in?

THE COMMONS?

The commons, or commonwealth, describes shared resources, often highly complex ecosystems, operating within what Wes Jackson would insist we call the ecosphere. The commons describes both the fabric of natural wealth relations that predates human contrivance and the systems we humans employ to govern our use of it. It infers the limits of the place as it changes in relation to surrounding places and events external. So, for instance, a semi-wild upland pasture and woodland system used in summertime by transhumant graziers for their sheep, goats, or cows could be considered a commons, as would the association of graziers and the procedures of their annual meeting and processions. In case you think these systems are long gone, I refer you to a well-timed email from David Bollier letting me know that still today, 21 million acres in Europe are managed as a commons, mostly for grazing.[1]

The Nobel Prize–winning academic Elinor Ostrum spent years of her life exploring and examining stable, commons-based governance systems in traditional economies around the world. From this evidence she set forth a set of rules that characterize these functional commons. These are the operating programs for durable, long-standing human settlements, and they play by certain rules and with certain recognizable patterns. She documented these formal and ritualized procedures in economic terms and helped interpret why such systems were able, over long periods, to accommodate fluctuations in the ecosystems, negotiate conflicts among users, and adapt practices according to technologies. (Read more in last year's *Almanac*.) Her rules for commons are:

8 Principles for Managing a Commons

1. Define clear group boundaries.

2. Match rules governing use of common goods to local needs and conditions.

3. Ensure that those affected by the rules can participate in modifying the rules.

4. Make sure the rule-making rights of community members are respected by outside authorities.

5. Develop a system, carried out by community members, for monitoring members' behavior.

6. Use graduated sanctions for rule violators.

7. Provide accessible, low-cost means for dispute resolution.

8. Build responsibility for governing the

Transhumant procession

common resource in nested tiers from the lowest level up to the entire interconnected system.

ABOUT THIS ESSAY

Obviously we live in a world of trade, and have lived in a fossil-powered growth economy for the past few hundred years, but we also know that there are real constraints up ahead, and that the stability, peace, autonomy, and identity of our home place requires that we meet many more of our needs within a bounded geography. That we import less, produce more, and despoil as little as possible. Fulfilling such a sensible mandate will require reorientation of current systems, creation of new and renewed infrastructures. It will require changes in settlement patterns and land-use, behavior, and allegiance. It will require patient social processes and opportunism on every scale, and unfortunately the social, political, and ecological climate is not particularly stable for the needed maneuvering. Martin Luther King Jr. wrote in the "Letter from a Birmingham Jail" that "all men are caught in an inescapable network of mutuality, tied in a single garment of destiny. Whatever affects one directly, affects all indirectly. I can never be what I ought to be until you are what you ought to be, and you can never be what you ought to be until I am what I ought to be." I believe in and aspire to live in some kind of federated commons. In this essay I propose a three bonding agents that set up a charged tension, within which what Wes Jackson calls "emergent properties" can emerge.

I include short, semi-moralistic micro-histories of landscapes I've been learning about in my hobo snowball way in order to orient the reader to the landscape logic that corresponds quite well with decentralist thinking and to approach the logic of commons-making within a landscape of theft, appropriation, and extraction. My themes are boom, water, and dust. Along the way I refer to sources for further reading and inquiry, so think of this as a temptation reel, not the whole story.

THE UNITED STATES LAND COMMONS

There are four federal public land systems: the National Forest System, the National System of Public Lands managed by the Bureau of Land Management (also known as BLM lands), the National Park System, and the National Wildlife Refuge System. The National Forest System contains 193 million acres. The National System of Public Lands contains 245 million surface acres, as well as 700 million subsurface acres, mostly in twelve Western states. The National Park System contains 84 million acres, with 57 million acres of it in Alaska. The National Wildlife Refuge System contains 96 million acres, with 76.8 million acres of it in Alaska. The National Wilderness Preservation System contains almost 110 million acres, 58 million acres of it in Alaska, within all four federal land systems. For more information, see www.wilderness.net.

It's a lot. These are the lands, landscapes, and ecologies that, though underdefended, still predominate our nation's surface. Technically these lands belong to and concern all of us— especially when it comes to issues of oil

ACEQUIA

Acequias are one of many ancient ditch irrigation systems in the world. The acequias in New Mexico are more than four hundred years old and made a hostile desert in the southwest habitable for the settlers who arrived up the Rio Grande corridor from Mexico. Prior to the acequias, indigenous Ancestral Puebloan farmers cultivated the uplands with small grid gardens. I'm sure they worked fewer hours than the colonizers did. It's quite a marvel how durable these hand-dug Acequia ditch irrigation system have proven to be. Unlike the New England small-town farm economy, it is still relatively intact. The water flows, the land is cultivated, while much of the land is in hay and forage. Subsistence gardens are still grown, and the traditional governance persists to this day. The acequias themselves look like small dirt canals, moving water from the central river up into the valley, slightly off-contour in a vague fishbone structure, allowing the shoulders of the valley floor to be watered—

and then in the desague, the canals return their water to the river.

Once a year in a ritual called *limpia*, the *parciantes*, or ditch members, all turn up with shovels to clean up the ditch under the direction of the majordomo. These ditches run the power

> You can see a nice film about the *limpia* process on archive.org in Robert Redford's *The Milagro Beanfield War*.

of gravity at an angle off the main river. This diversion ditch spends out its water according to a precise schedule overseen by the majordomo, watering the crops of the various users. The water leaks from the ditch and seeps through the ditch walls, supporting cottonwood trees, bushes, and birds. When released with the little metal gate, it floods briskly across farm fields.

---

extraction, grazing regulation, mining permits, and pipeline development—with profound climate implications for all of these activities. It is estimated that 23 percent of total U.S. carbon emissions come from public lands mining and gas permits. According to advocates from the Rainforest Action Network, canceling these permits would reduce carbon emissions by 450 billion tons.

The federal government owns 47 percent of the eleven Western states.2 These lands are currently administered thru the state via our federal agencies. Unfortunately, armed militants, a growing group of Republican senators, and a suite of Koch brothers–funded front groups, such as the American Lands Council, propose to weaken the protection on these lands. They wish to proceduralize a massive transfer ownership and management to lower state agencies, loosen rules, and issue more permits for more mining, ranching, and other economic activities on these lands. In 2016 a group of rogue armed ranchers led by the Bundy brothers seized and occupied a federal wildlife refuge, set up to protect migratory wildlife—undoubtedly a commons, claiming that this land should be under local control, and therefore more accessible to people like them. White, armed

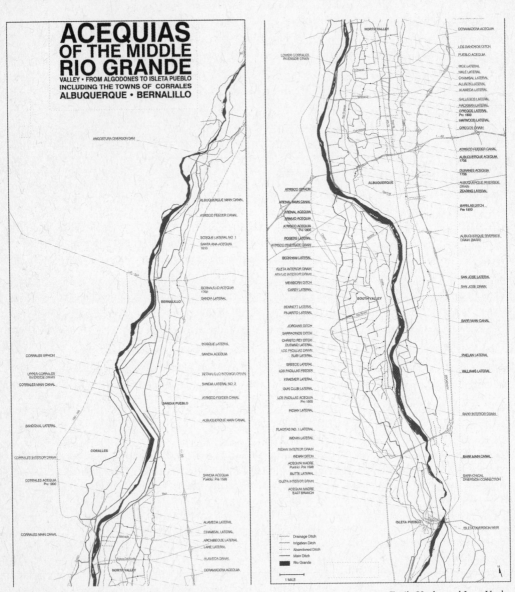

Field Guide to the Aceqias of the Middle Rio Grande

Emily Vogler and Jesse Vogler

What does not evaporate slowly rejoins the mother river through temporary rills and gutters created by the shovels of the farmers once the gate opens and releases the flow. This right to irrigate comes in intervals, coordinated by the majordomo, a manager hired by an elected board, a highly democratic committee that oversees the administration of the ditch system, including its cleaning and inevitable conflicts. When there is not enough water, everyone gets less water: fewer hours when the head of their gate is open. When there is even less water, acequias must negotiate among themselves to share the water of the river. In extreme drought this can mean only one hour of water every ten days. The negotiations follow the landscape, shaped by the contour. Relations are fixed upstream and downstream. The system is predicated on finding a truce, getting along. Predictably, some families come to have more power than others, though this is relative; in times of drought the water is shared equally. The skills of ditch management include accommodating,

finding harmony tempered with obligation, and living with constraints. As such, the Acequia system is a powerful model of commons-based governance. Read more about water appropriation at www.waterencyclopedia.com/Po-Re/Prior-Appropriation.html#ixzz4RGZyS9Yg. Learn more about acequias from the videotaped sessions of the Our Land[2] Symposium at www.agrariantrust.org/2016Symposium.

BOOM SUCKA: THE GOLD RUSH

In 1846, San Francisco Harbor wasn't too busy. According to the harbormaster, in that year it received nine whalers, sailboats carrying sailors from New Bedford, Salem, and Nantucket to spear and liquify the sperm whales, boiling down the fat into barrels aboard ships. These sailboats were outfitted with meat and provisions in California and went on their way to hunt the ocean for fat. Only three merchants ships came through the port that year. Only two years later, in 1848, gold was discovered in California, and by

Utah already passed the Transfer of Public Lands Act, setting up a massive land grab for states struggling with budget shortfalls, who are quite likely to enclose and liquidate these "assets" as they start to bear the enormous costs of stewardship and administration currently undertaken by federal agencies. It is very true that public lands managers and agencies are chronically underfunded and overstretched—they have too much acreage to take care of now, but that doesn't mean the solution is turning the lands over to market forces.

In academic circles, commons are conceptualized as community-based systems managed outside the state, outside the marketplace, usually embedded in the cultural land-use patterns of long-settled areas. These practices arise in place and underlie the sovereignty of the people

the next year there were 775 ships that had arrived in San Francisco, bringing miners, speculators, shopkeepers, supplies, munitions, picks, and shovels for the front of discovery. Fortunes were made. The steamer *California* arrived in San Francisco in February of 1849 via the straits of Magellan; by 1852 it had shipped $122 million worth of gold. California gave a powerful boost to American shipping, and shipyards all along the eastern seaboard were kept busy. Cargo moved briskly and at boom prices, setting up and normalizing labor inequities on the West Coast that haunts us to this day.

Records were set in the years that followed on the passage from New York or Boston to San Francisco—eighteen vessels made the voyage in under one hundred days. Those boats were sailing two to four hundred miles per day. Many of those who sailed to San Francisco during this time were not sailors and wanted only transport to the mines of California; given the chance, they abandoned ship right away. The captains, anxious to dispatch their cargoes in haste, would call for "men along the shore" to come and unload. The frequent and unscheduled arrivals of so many boats created favorable conditions for the longshoremen, stevedores, and warehouse workers to organize for better pay and working conditions. They formed a union and eventually brought forth an incredible set of institutions, including the hiring hall or dispatch center, a physical building where union members would go to find employment. Eventually San Francisco longshoreman even created subsidized housing. But during the boom, to get out of harbor, the captains, traders, and ship owners would often shanghai a crew, getting men drunk and hauling them aboard in a stupor.

Soon enough California became a state of the Union, robbing Mexico of its Spanish inheritance, the long-tended Mission outposts, and was connected by Congressional subsidy to the East Coast by a ten-day overland route from Missouri to Sacramento, getting news and transacting business across the continent far faster than the ship's passage around Cape

---

who live there. Due to the displacement of most indigenous communities, it is hard to find very many fully functioning models on the North American continent, as most settlement patterns were predicated on extraction of commodities for foreign markets, so quite few of us are familiar with the patterns of commoning so well practiced on other continents. Some exceptions include the manomin or wild rice tradition.

Land is disjoined and noncontinuous. Land takes so many shapes, has such a various temperament. It is hard to understand land as a commons, hard to see beyond all the dividing and changing, the making of fields from forests. Water has the advantage of being obviously fluid, connected to itself, all connected together—it may be our aquatic avenue for this discourse on the commons. The ocean is so hard to see into. We feel it like a surface, sometimes rough sometimes smooth and undulating. It's hard to comprehend as a dimensional volume, with mountain ranges, chasms, and deep currents welling up.

The atmosphere could be called a big commons, as could the universe. But in considering or administering the health of a commons it is easier and more practical to deal at a smaller scale: a scale where decisions can

Horn. The Pony Express ran a relay-race of fast riders, a stopgap until telegraphs could be run. Transferring a mail sack from hand to hand, each rider worked a specific hundred-mile segment of the trail, with fresh horses stabled at stations every ten to fifteen miles along the way. Buffalo Bill allegedly got his start as one of the more expert riders, outrunning not only steamships and packets but also the frequent attacks by the indigenous people, especially the Paiutes, over whose territory they crossed. According to legend, Buffalo Bill once rode more than 380 miles without resting for more than four hours. Fact-checking reveals that this feat was actually accomplished by a man called Bob Hallam in 40 hours. Listen to the whole "Story of the Pony Express" on Librivox. To learn more about the Missions, read *Life in a California Mission: Monterey in 1786* by Jean François de Galaup, compte de Lapérouse.

Miners, prospectors, and speculators soon intensified their methods. Sluiceways became hydraulic sprayers, forcing high-pressure water in hoses to dislodge the riverbanks, washing whole riparian habitats down into the delta along with the mercury and arsenic. The need for water to loosen the veins of metal from the mountain created the legal architecture of prior appropriation: "first in use, first in right," which has come to define Western water rights since that time. Ingenuity, opportunism, and quick cash built up a supply chain behind them as they carved into the mountains. One man took care to purchase every pick, shovel, and pan before running through the streets with gold dust to precipitate the frenzy. The wealth of California built a market, skills, appetite, and conveyance that was needed to profitably extract other minerals as well, borax and talc powder from the great desert basins, as well as silver, tin, and finally oil. Imagine twenty-mule teams hauling washing power from nineteenth-century Death Valley (see www.mojaveproject.org). You can still visit these lonesome small towns perched on sulfurous springs, connected by rail to a market for their dust. The railroad is gone now, but the

---

be made by people and not by computational algorithms. On this scale, commons are bounded by geography and relation. To participate in one is highly specific and contingent. For example, the acequia system of the southwestern United States and the highly regimented grazing territories of European upland pastoralists are both living and functional commons-based systems. Commons have rules, and those rules relate to the rules of nature. As Oren Lyons says, nature's rules have no mercy. Human commons management therefore must accommodate those rules and provide governance over egress, scale, rights, responsibilities, bonds, and the boundaries. Because commons are natural systems, we humans are not totally in control. Across geographies and histories, human people and human communities have bonded in shared commitment to place—they have formalized their sharing in communitarian land-ways with strict rules. These rules are hardly known to us, enmeshed as we are in the current predominant "private property" worldview. In a commons, one does not have the right to destroy or despoil one's portion of the land. It is not ours alone; it is shared. The theorist David Bollier described "commoning" as the set of behaviors, rules, and cultural assumptions that it takes to govern ourselves

Tecopa hot springs, raucous wildflowers, and date oasis make for a fun trip.

The gold rushes of 1849 and of 1859 were powered by individuals and small posses working their claims. Artisanal miners, they would be called today, mining with minimal use of mechanical aids. In 2016 there are an estimated 25 to 30 million artisanal miners in the world, critical producers of minerals for the digital economy. The scars left by 1850s gold mining may be cruel, but compared with the scale and toxicity of the wounds bleeding now, it's almost a quaint injury. Would that more of us would notice that we're walking in shade of small canyons created by wagon wheels, rain, wind, and time. The eroded gully is a haunting, unintended consequence of a racing, roaring current of human agency, racing down into the earth with their little picks, making it into money, as mud washes down to the ocean.

HEADWATERS

The Richest Land is a short film created in 1972 by the Land for People Coalition. Its aim was to examine the political economy of state-funded irrigation projects and specifically, to call attention to the nonenforcement of acreage limitations in the Reclamation Act of 1902. This small band of advocates made a clear and well-reasoned argument all the way up to the Supreme Court for a fair playing field between large corporate operations and smaller family farms. They effectively exposed all the subsidies, cheating, and hidden tax incentives that benefit agribusiness syndicates on top of the massive taxpayer-created irrigation infrastructures that water the fields. The film starts its narrative up in the Sierra Snowpack of California, where so-called "white gold" (snow) makes possible the millions and billions of dollars of production down below. According to their film, the value to agriculture of the annual snowpack in California was more than the total value of all the gold ever mined. That white gold is channeled down through dams, pipes, pumps, and canals. Although homestead rules and federal rules legally limited farm size to 160 acres, this rule was ignored by speculators and agents who conspired to build a fabric of megafarms. See Grey

---

within a bounded and ecological commons. Commons require discernment, belonging. Andrea Nightingale writes that social factors are prerequisite to successful commons. She says that a sense of "group belonging," of feeling certainty in a shared and collective identity, is fundamental to commoning.

Commoning gives us a language to celebrate what we share, what we rely on others to access, and what has not yet been enclosed. Greenhorns celebrates commoning practices because we recognize that our movement needs territory. The farmland we'd like to farm is a commonwealth, whatever its legal designation, and belongs equally to our children and grandchildren as to us or its current owners and, one could argue, equally to the animals who coinhabit it. If we're serious about sustainability as a culture, we'd better borrow it in ways that allow the land to live. We'd better get busy learning how to belong to the commons we're building, even as the ecosystems we inhabit become ghosts of themselves.

Brechen, *Farewell, Promised Land: Waking from the California Dream* (Berkeley, CA: University of California Press, 1999). Piped and pumped, damned and dripped, sprayed and siphoned across the big Golden State. The large landlords ignored the rules, and when finally challenged in court, had the legal size increased to 960 acres. The siphons of justice couldn't refuse the powerful vacuum left in the pipe.

The lettuces are shipped out in wagons by refrigerated rail, and soon the workers are shipped in: braceros living in warehouse barracks, often right along the tracks. Mined water, mined farming, mined labor. This destiny did not go unchallenged; struggles to force California to temper itself make for a great subaltern wormhole. Dorothea Lange's husband, Paul Taylor, the University of California agricultural economist, spent his career struggling to confine the unchecked power of agribusiness to make rules more compatible with human values. He did not succeed, and at the end of the New Deal, which he spent documenting in *On the Ground in the Thirties* the tragic conditions of farm labor camps, he brought his aspirations abroad to the Philippines, Vietnam, and Pakistan,

A picture of Paul Taylor, his droopy eyelids taped up for the Land and People Conference on Land Reform.

implementing land reform projects for the State Department to stabilize economies during the buildup to the Cold War. The state department understood that cultural stability and economic stability were improved by the more egalitarian political economy of this kind of farming.

Walter Goldschmidt wrote an incredible study for the U.S. Department of Agriculture showing the social scars created by absentee ownership and farm scale. He tracked childhood mortality, civic life, church attendance, quality of roads, nutrition, and voter turnout; all fared markedly better where farms were smaller and operated by their owners. And yet, these well-researched and well-intended efforts were largely in vain. Water barons and speculators got their way, stole water (see the film *Paya*, about the Owens Valley water grab). Today, the Resnick's family, of Los Angeles art world fame, hold title to a metaphorical Golden State center pivot—the publicly funded, now privately owned Kern Water Bank, right along the spine of constructed megawater infrastructure. A senior attorney at the Center for Food Safety, calls it "an unconstitutional rip-off." The Resnicks make make millions a year selling water back to the state and other water users, as they control multiple vertebrae of state-funded infrastructure and use more water to irrigate their fields than many cities combined.

The Reclamation Act of 1902 paid for dams by selling off public land to raise capital for all the cement and engineering needed, once again mining the commons to concentrate profit and power. Eventually we dammed nearly every major river in the West. The Gila River, which flows from the first national wilderness area created by Aldo Leopold in New Mexico, remains one of the last free-flowing rivers, but it currently faces its fourth damn proposal—an extension of the Arizona CAP canal project. Learn more at http://ca.statewater .org and the Gila River Resources Project. During a lecture by one of its defenders, I learned from a passionate environmentalist in the audience that one of the proposals for use of the water is for Facebook's server facility—a multibillion-dollar set of warehoused computers that will hold aloft all our digital likes, dislikes, links, and friendships. According to Greenpeace, if data centers were a country, they would rank fifth in use of energy.[3] The Internet is the largest thing that humanity has built. It requires electricity, chemically purified water, and plenty of other invisibles.[4]

If Silicon Valley is leading us off a cliff like lemmings, can we contradict its logic in an offline forum? California is widely regarded our nation's agricultural and technological powerhouse. The bullshit prosperity comes from a succession of booms, programs of domination from stagecoach to syn-biotech. The Central Valley is literally sinking from subsidence, the earth literally saturated with irrigation salts, the cost of living the highest in the country. Chinese investment in

Hollywood, Multibillion-dollar water bonds— California, you are a Potemkin Emerald City. I will not be fooled!

Oh, California, this golden bear with her plump tummy—her expensive velvet gown of lettuces, asparagus, almonds, and citrus trees. She affords it because of a little ol' conspiracy made from cheap land energy, mega-engineering, federal subsidies, and crony capitalism. Will the long-awaited quake spill out her too many residents across the deserts into Albuquerque, Flagstaff, Tucson? Will she be forced convert her fields from twinkling green back to beige and brown of limas, wheat, and drylands agriculture? Can we in other regions speculate on California's contraction and take heart that the produce of other soils, the beef of wetter prairies, the rice and cotton of less engineered landscapes, will come out ahead in the end?

California may drive the prices, standards, quality, and work-speed for U.S. produce and specialty crops, but it's the one with 70 percent of its workforce undocumented and unprotected. It's the one with an automatic pest invasion control mandate to suppress any invertebrate insurrections against its precious precarious monocultures. It's the one that designs the ear buds, but how can it possibly last? How can we believe in this obviously failing dream? She sucks harder from groundwater and reservoirs. She pumps and pumps squeakily. Wells drop, new wells are sunk, reservoirs drop. The precipitation just doesn't come. Respected thoughtful farmers

say, "We'll see about next year when it comes."
It doesn't seem like a plan to me.

ORANGE YOU GLAD THERE'S A PLAN?

Eliza Tibbets was a Sweedeborgian, spiritualist, feminist, and suffragette. She came to Southern California, like many other educated Easterners and prosperous Midwesterners, for her health. Good air, they said, would cure you. It would probably cure many of today's diseases also. Eliza liked it fine. Through some kind of premonition arising from good health, or maybe spiritual insight, she surmised that oranges would grow in Southern California, and resolved to give it a try. In 1873, Eliza wrote a letter to her former neighbor, now head of the newly formed U.S. Department of Agriculture, George Saunders, in Washington, D.C., and received a few young trees. (Saunders, an avid fruit explorer, was also a cofounder of the Grange movement.) The trees had come from a monastery in Brazil from a spontaneous branch mutation or sport that grew on one of these monastic trees. These were not the little seedy oranges of the padres, nor the lemon-stock they had brought with them from Mediterranean; these were New World fruits, big and juicy, evolved in the outer orbits of empire, under the hot breath of Bahia. She watered the new trees dutifully with her dishwater, and they flourished. Thus was born the Washington Navel boom of Southern California.

Eventually, Washington Navels grew on more than 20,000 acres. By 1920 the cooperative packing, shipping, marketing, and product

organ, the Southern California Fruit Growers Exchange, later Sunkist, was averaging $100 million a year from the sale of citrus, which included the navel oranges. What a delightful fruit: large, plump, bright, beautiful, smooth skin with a rumpled button on one end—and thirsty. Easy to ship, juicy, and long-lasting. This was the orange that launched a land boom. More and more health-seeking small farmers came onboard with tidy ten- to twenty-acre parcels and sweet Victorian-style homesteads. Many of these new émigrés to the scenic valleys of what is now known as the Inland Empire were highly progressive and well educated. They build lovely homes and planted their streets with carob trees—fodder for goats—and flowering jacarandas, fueling a real estate boom. According

a USDA Research Experiment station. It was a cooperative, formed on noble principles, with multiple intersecting spheres of democratic decision making, with diagrams of relational accountability and a whole series of institutions created for mutual aid. Like the first co-op, founded by industrial workers to get cheap bulk tea and sugar from plantations of the British Empire, it is a form adapted to particular purposes of its stakeholders. From a commons-governance perspective, the critique would be clear: it lacks accountability to its ecosystem context, is bound up in a global commodity chain, and the actual land workers are not part of decision making, so they cannot transmit their insight. Predictably, therefore, as a social form it became quite stagnant, homogeneous, and nonadaptive—a convenient structure that allowed predominantly white owning-class operators to contract out the various orchard management tasks, not thinking very much about what comes next, to its peril.

to the history books, Riverside became the richest city per capita in the country in 1880.

The pattern proceeded in this order: immigration, speculation, irrigation, idyllization, cooperation, transportation, supplementation by petroleum. We could make an argument that the relative prosperity of this short period of golden idyll had only marginally to do with the basic economics of oranges, and far more to do with external factors. Still, the oranges needed to be picked, washed, packed, marketed, and sold to distant markets. Sunkist sold the oranges, the landscape, the lifestyle, and the appeal, accelerating the boom. Sunkist coordinated the efforts of growers in putting together irrigation districts, controlling pests by raising and deploying biological controls, and got themselves

Today's California citrus industry is contending with convergent serious threats, with cheaper Brazilian fruit high upon the list. Sunkist's super-PAC lobby group is working to defend blind exterminationist frenzy as descendants of the citrus junta to spray with impunity poisons made by Bayer. These are neonicotinoids, or neonics, a class of nerve-impacting insecticides used as soil drench that purports to control Asian citrus psyllid (ACP), the vector of citrus greening disease, a fatal ailment for orange trees. What we know is that Bayer's imidacloprid kills bees, kills birds—kills up and down the food web. In Santa

Barbara County last year every single stream was found to be contaminated with it. And so the monocultures of commodity citrus, the expensive dress worn by these desert valleys, will be sprayed and fumigated and picked for juice by hands imported from over the mountains. The groves are managed by Grove Care franchises who prune and spray and fix the leaking irrigation when needed. This prime real estate is now tucked and dotted with Hollywood ranchettes, only a few of whom are considering the future of this agricultural landscape, dreaming of stone fruit, or olives, pomegranates, carob, oaks, or the

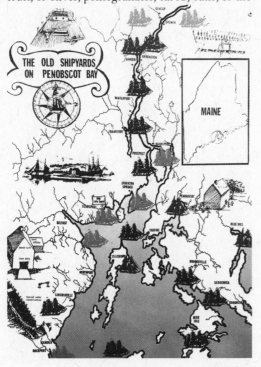

other fruits more suitable to the rainfall. At least not yet.

PORTS

Up near Canada, way down east in Maine, is the now-extinct fishing economy of Eastport and Jonesport. It's a brick town, and some mansions remain; these were fish fortunes evolved up from the weirs of the Passamaquoddy to sailboats deploying little dory boats, men hauling miles of wet line, coiling it into buckets on deck. Yet another tragedy of the commons. In 1880 there was a steamer to Boston five days a week. The shores bustled with sardine trade, canneries, and the still-running mustard mill for packing fish. Sardine speculators built model factories and brought in fleets of Norwegian women, known for their deft fingers. Across the great bay wafted the smell of smoked fish. Puss in Boots, a cat food factory, was one of the last to close. Now, though the highways are loud, dented by trucks hauling gravel, blueberries, bait fish, Christmas wreaths, and baby eels, this easternmost city in the country doesn't even have bus service. Read about this fascinating town in the *Quoddy Tides* newspaper.

Farther south, in midcoast Maine, Bath, and the Kennebec region, the fortunes came from shipbuilding and what the ships carried. Proud prominent prow of a white clapboard mansion with a widow's peak and black glossy shutters, lilacs in the dooryard, and within, the women who held the wind: the women whose husbands built the ships, sailed and fitted them. These captains and sailors and traders who discovered and invented trade routes, improvised new ones, dashed out ahead of embargoes, privateers, pirates, and hellfire without engines, without email, without any weather forecasting. Some women went on board with their husbands to trade in tea, pepper, lumber, sugar, and molasses, mounting their kids on deck in a kind of nautical highchair, doing laundry at sea. Some stayed home and in their apron-skirts held together the civic life of their towns. These houses were beneficiaries of frothy profit from California's goldfields, from slavery in the East Indies, from the smuggling trade. Even farther south, in Salem, the mariners created a fellowship for mutual aid, a seaman's society: the East India Marine society, a seamen's pool to provide for indigent sailors, that money wrestled from the wind.

THE GRANGE

Looking out over the harbor stands a proud little edifice of extraction: the Grange Hall, the Odd Fellows Hall, the Knights of Pythia, the Knights of Columbus— or one of dozens of fraternal orders created by working people of land and sea in the nineteenth century. These days, their limp velvet pomp has faded, the sweetly fragranced closet of robes, sashes, choir books, and carefully labeled regalia, if it is preserved at all, by stubborn octogenarians. These places represent a powerful infrastructure for community gatherings, pot lucking, political agitation, cooperation, other rural organizing activities. As young grangers, young farmers, rural dwellers, many of us

reflexively celebrate the populist vision of our grandmothers. We learn and sing their songs alongside the grange-hall piano. We wear the sashes over our lumber-jack suspenders and slap our knees with nostalgic excitement, a feeling of belonging. Greenhorns even released a vinyl record of grange songs, *Brian Dewan Sings Songs of the Patrons of Husbandry*, a project of Grange Future. This rabble-rousing populist tradition suits us fine—a beautiful, hopeful, humane version of nation-building, a progressive moral farm family economy. Only recently dispossessed by enclosure in their source countries—Scotland, England, Ireland—they banded together to protect their yeoman-stake in the New World. And though it included women from the outset, the Grange excluded many others.[5]

DOWNSTREAM, THE DAUGHTERS OF EXTRACTION

As we, a bit more than a hundred years later, sit in their sunlit and musty seats, we remember ourselves as the daughters of extraction. These grange halls were built from the prime time of ecocide, when began the removal of 80 percent of the standing forests. These halls were built by the hands and boards of freshly felled forests, from the the fresh flesh of territory undermined, logged, and stripped. The proud architecture, the workingman identity crafted and hewn from the destruction of natural wealth, born and attuned to the sound of saws, the frying of doughnuts, the speedy deft splaying of salmon on racks in the sun. The fat of the land, the fat of the sea.

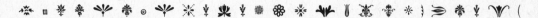

The workingman identity and pride of place, converting that place into a muddy settlement, now collapsed.

We are all implicated as descendants of this intertwined history of betrayal, conquest, decline, collapse, and surrender, either as beneficiaries or as the dispossessed. Daughters of extraction, and probably at least a half-breed daughter or granddaughter of expulsion, enclosure, dispossession, and betrayal. And now, in this politics of fear-mongering, wall-building, racism, misogyny, the victor-identity has got a knife it its mouth, overt in its plunder-lust, even if the true biography for most of us was of being plundered, at least on one side of the family.

We are inheritors of what Donna Haraway would call a "string figure"—the multidimensional knot that kids play with in cat's cradle—the loops and twists of rope in fingers. We are implicated; we accept the pattern into our hands by living. My own great-grandmother was able to race for a homestead in Oklahoma because she had no husband and claim that land, and settle it, but was later driven off by dust and violence. I read that in the Southwest, forcibly evacuated indigenous people were forced to abandon their small children, who couldn't walk the distance, to the households of early settlers. These formed mixed households of indigenous, Spanish, Muslims, and Jews, themselves escapees from the religious persecution of the Inquisition, all hunkered in the desert pretending to be Roman Catholic, avoiding scrutiny in the scrublands, making their own unique culture.

Haraway characterizes these fusions of cultures and destinies co-dwelling and co-becoming as "syn-poesis." The syn-culture got cooked up in small valleys between big mesas, made of quartz, with limited water: a concentrated experiment.

SAIL TRADE AND RESISTANCE

Each place has its own version of this same story, how empires washed ashore, a special part of that place was plucked as treasure, which people of that place were loaded involuntarily into the labors of lime kilns, sugar plantations, or enslaved to exported as sculptors and irrigators of new lands in the Carolinas. Sailboats arriving bringing sharp shovels, blades, picks, salt, commissary, bringing trinkets and buckets, ballast and biscuits, and inequality. Barricades, surveillance, militarization, fear—these are not new; they are only newly applied to those of us who won the previous rounds, joining in the legions of the already oppressed, the refugees hurling themselves into diasporas of desperation from global land grabs, river dammings, enclosures. People forget their land-places, traditions, skills, and seeds and become displaced. It has all been theorized, actualized, documented, and made into journalism for decades. We know this.

Trade is not incidental to colonization and conquest, the violence and overruling; it is the driver. Arriving by ship with guns ablaze, defending the hold under the stern, the great cavernous appetite. What violence danced out across the waves, elegant, with canvas aloft. Trade set the terms for foreign relations, its arbitrage

a calculus of improvisational undervaluing and overvaluing. Sail trade marks an inflection shaping each place it touched, carving new routes, as the empires danced their tactical chess games across mountain ranges and harbors as towers, knights, queens, and rooks moved across the coastlines, across the continents.

Arriving on a beach to negotiate with the indigenous kings for nutmeg, pepper, tea, opium, gold, and myrrh. Flattering the vanity of an Egyptian queen who sought trees to adorn her palace. The sail trade era is not just a harmless romantic epoch, a ballet in canvas along the shores of history; it brought global capitalism with it, a reorganizing factor that built the warehouses, ports, canals, palaces, and fortifications. The World Trade Center, the world trade agreements, the structural violence of NAFTA. For a readership of producers, it's a history obvious in its injustice (see http://www.voicesofmaiz.org/english).

SMUGGLERS

Here is a picture of the *Grayhound Lugger*. It was built over the course of two years by shipwrights and its captain, who now lives aboard with his family. Luggers like this one hovered and ducked around the Cornish coastline in the eighteenth century, smuggling whiskey, tea, and other goods. It was estimated that one-fourth of the horses in Cornwall were used for smuggling back then. This I learned from a lovely little monograph book on the subject, *The Cornish Smuggling Industry*, which I highly recommend.

Smuggling is another way to characterize trade outside empire's artificial constraints. These people called themselves free traders as their competitive advantage, their USP—unique selling point—was cheaper prices, making tea and liquor available to the working man. They weren't employed by price-inflating monopolies, nor did they deal with the bureaucracy in the customs houses that imposed their fees, taxes, and tariffs. No, they were direct traders of contraband—responsive, nimble, opportunistic, with a high level of social custom compelled by their craft—getting on and off the beach by the light of the moon. The commons of the seashore makes available this kind of horizontal, spontaneous, and extralegal trade.

The ocean commons allowed for autonomous

parallel trade routes, both sanctioned and unsanctioned. Yes, there are harbors with forts and fussy bureaucrats, but there are also inlets, rivers, and peninsulas to smuggle across. Very few geographies are totally governable.[6]

What insights can we draw from the understory of trade that can guide us in reconfiguring an economy more logically adapted to the underlying landscape? We see the sunken dock, the port, the half empty warehouses, but these do not teach us what we need to learn.

What we need to learn is happening in the understory: the understory that sustains the trade, and may be defiled, but is not defined by it. The water, the gold, the citrus, the river, the fishery; the commons underneath, the rules of nature.

FORESHORE

It all washes down to the ocean, and the nutrient finds its way into the bodies of the seaweed, for better or worse. Limpets, crabs, urchins, mussels, clams, sea stars, whelks, periwinkles, barnacles, anemones. Upwellings, phytoplankton, and the earth-magnetic currents penetrating aqueous depths with life. The seaweed is mother shelter to them all. And the fat prosperous gulls— of the family Laridae, "ravenous sea bird" in Latin, filling their gullets with live and dead and rotting alike. The seaweed threshold, a welcome mat, washed up on shore in a line, like a smelly eyebrow, pierced and polluted and overplucked, above the edge of the liquid ocean. A holder of flotsam and junk. A nest for beach flies and smelly bits. The seaweed holds its own lessons. According to Donna Haraway, mutability and responsiveness are the crux, the clutch, the core tactic of life systems. This is its metabolizing responsibility, which she calls "response-ability," the agency to react, to respond, to adapt. For the victims of whatever empire it has meant making meaning and making do, either in the margins, in the fall out, or in the debris. Donna points to the many rubbings and interfaces, interstices and touchings between forces and beings, all sorts of rearrangements that yielded successful outcomes. Organisms that were partially eaten

SUMMER 2017
The Greenhorns is working to support Future Farmers, an art collective, in another sailboat art stunt, this time carrying ancient grains from across Europe and the Mediterranean basin back to Palestine, celebrating the diaspora of these grains and the breeders and places that influenced and unfurled the genetic potential of the seeds. Learn more at futurefarmers.com/seedjourney.

SAIL PLANS of the CLIPPERS

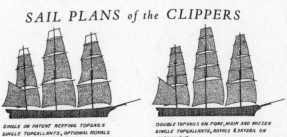

SINGLE OR PATENT REEFING TOPSAILS
SINGLE TOPGALLANTS, OPTIONAL ROYALS
SOMETIME BARQUE RIG WITH NO SQUARE
SAILS ON MIZZEN MAST

DOUBLE TOPSAILS ON FORE, MAIN AND MIZZEN
SINGLE TOPGALLANTS, ROYALS & SKYSAIL ON
MAINMAST
MIZZENMAST OFTEN RETAINED SINGLE TOPSAIL

but not digested by other creatures became the mitochondria inside those creatures. On multiple scales, the evolutionary soup sloshed in and outside of categories quite wildly, creating as a consequence diverse and beautiful ecosystem of creatures and cultures. It is in these mix-ups and mash-togethers, in the absorbative downstream commons, that I go burrowing for insight on how to cope.

How do we go about interpreting these teaching places, teaching systems, teaching commons? When we look at land commons traditions, including those that inspired Karl Marx, Leo Tolstoy, and Henry George, we find that many of their theories derive from the distribution of marginal lands—bypass systems operating according to different rules. Unsanctioned, marginal, waste, backwoods areas acted as sanctuary and buffer from the starvation wages and exploitation of the dominant culture. These outer places, higher slopes, wastes, moors, and forests act therefore as an outlet, subsistence, and basis for long-standing resistance to the central power—a totally different set of central metaphors. The human systems, governance,

and administration of these places and systems emerge therefore as a countervailance to the impulses of empire. Source of firewood, ritual, banditry—these are places and systems that have evolved for cultural survival, a kind of smuggler logic, persistent understory in the genetics of dominant culture. How do we become more intimate with it?

In Brittany I met farmers, bakers, and millers working together with land occupiers at Zone à Défendre (ZAD). They see themselves as part of a global resistance movement; they name themselves with an electrical metaphor—a short-circuit bypass to the corporate control and lobbies who want mega-airports. In bypassing the corporate food system, these foods and materials can flow in solidarious and local markets—reciprocal exchange socially and informationally dense. Empires may continue to dominates the rich river valleys, the main ports and harbors, mine our mountain tops and maneuver four billion shipping containers across the oceans, but in small harbors and civic plazas, under tents, among friends, up rivers, in churches and school basements, with basic coordination and good

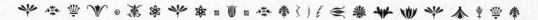

social skills—we smugglers are unvanquishable.

Probably it will take what we're already focused on: direct action at the scale of humans, families, communities, and extended networks. A more diverse, prosperous, fair, sustainable landscape seems still the right goal, but I am starting to change my notion of how that possibility will be transacted. More as hand-to-hand motion, like the acequia ditch, rather than administered by an agency, a conservancy, or a sanctioned bureaucracy. Add to this the stark demographic truth that these tasks are on our generation's shoulders. We, our friends and friends of friends, are yoked to this project. The gray-hairs have social security to look forward to. Our personal actions, intersections, and improvisations— based in a landscape basin of relationship, commitment, and stubbornness.

COMMONS REPAIR

FDR enrolled 3.4 million men in the Civilian Conservation Corps, which built 13,000 miles of trails, planted more than two billion trees, and paved 125,000 miles of roads. As Aldo Leopold said, the president failed to resolve the "standard paradox of the twentieth century: our tools are better than we are, and grow better faster than we do."

The evolution from a simple set of judicially created rules to a statutory permit system reflected the transition of the West from a livestock grazing, mining, and dry farming economy to an increasingly large-scale irrigation society with urban oases supported by aqueducts and three multipurpose dams providing carryover storage and hydroelectric power. Prior appropriation, especially as it was incorporated into the law of equitable apportionment, permitted the storage of water in progressively larger carry-over storage reservoirs and thus allowed the Western states to buffer themselves against both chronic aridity and the cycles of rain and drought that plague other parts of the region. The state's interest in water allocation increased as the West grew, but the legal impact of this evolution was muted during most of the twentieth century. If one were to simply read the cases, prior appropriation progress would merely be from an underdeveloped to a mature body of law as reflected in the great early twentieth century Kinney and Wiel treatises. But, the focus on the formal law ignores the significant changes in its function.

This will be hard work. Much of the terrain currently occupied by the organic sector was converted from family farm configurations, not plantations. A lot of the countercultural terrain was carved from indigenous habitats, upper watersheds still springing with juiciness. Organic Valley, though it exports its products around the country and produces on half a million acres of farmland, is based in the small, winding, riverine Driftless region, a set of valleys inhospitable to monoculture. The land base for organic agriculture was converted from homesteads of subsistence economies and infrastructures. Tackling the abused bottomlands of empire is going to require serious conviction and likely some subsidies. Which means that land repair is going to need some powerful champions.

According to Slow Food, 70 percent of U.S. farmland is used to grow feed for animals, and that is the land that needs repairing. According to conservative estimates from the Royal Agricultural Society, England's soil has only a hundred more harvests. John Jeavons and Wes Jackson have been talking about this for a long time. If there is anything crazier than chasing whales in the ocean, surely it is replacing the world's most complex ecosystems to make palm oil for crunchy packaged cookies and agrifuels. Can the current fads for top-down techno-fixes give way to a bottom-up empowerment mode? Wall Street bankers—with their research on mylar pouch-food trends for the outward and mobile—are speculating on date palm orchards in Indio, where the Colorado river is pumped across the desert. In Arizona they are sinking wells in a land boom for alfalfa and pecans. In the Willamette Valley it's a blueberry and hazelnut boom, with eyes cast at Asian superfood markets. Today's agricultural players are using metadata from Goldman Sachs.

This land needs repair; denied the revivifying cascades of animal urine and passing through years as pasture, these lands are very obviously dying. Will the robots consent to roll over them for a paltry return? As crop failures leave bald spots and holes that need filling, will some of these fields become opportunities that may be considered worthwhile for rehabilitation? Will some billionaire set a bounty on land repair? Will these carbon-farm narratives create an opportunity for young bravery like what's happening at the Savanna Institute (www.savannainstitute.org), where they're converting soybean fields into agroforestry? Or our friends at Mastadon Farms and Versaland Farms, who are putting in chestnuts, hazelnuts, walnuts, apples, plums, cherries, and oaks by the thousands into the stale stubble of GMO cornfields? What makes this more possible? Will it be these reformed bankers and landed big shots with "Save the World" regenerative agriculture narratives? Let's hope so!

NOTES

1. Eurostat, "Common Land Statistics," http://ec.europa.eu/eurostat/statistics-explained/index.php/Common_land_statistics_-_background.
2. Congressional Research Service, "Federal Land Ownership: Overview and Data," December 29, 2014, https://fas.org/sgp/crs/misc/R42346.pdf.
3. Gary Cook, "How Clean Is Your Cloud?" April 2012, http://www.greenpeace.org/international/Global/international/publications/climate/2012/iCoal/HowCleanisYourCloud.pdf.
4. Kate Singer, "An Electronic Silent Spring," 2014, http://www.electronicsilentspring.com/primers/energy/.
5. Learn more from the songs of Robin Grey, www.threeacresandacow.uk.org.
6. See James Scott, *The Art of Not Being Governed: An Anarchist History of Upland Southeast Asia* (New Haven: Yale University Press, 2009).

CALENDARS AND DATA

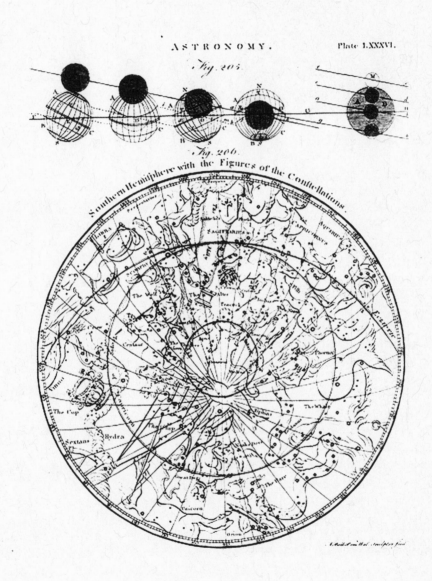

ASTRONOMY.

Fig. 205.

Plate LXXXVI.

Fig. 206.

Southern Hemisphere with the Figures of the Constellations.

PHASES OF THE MOON

FOR THE YEAR 2017

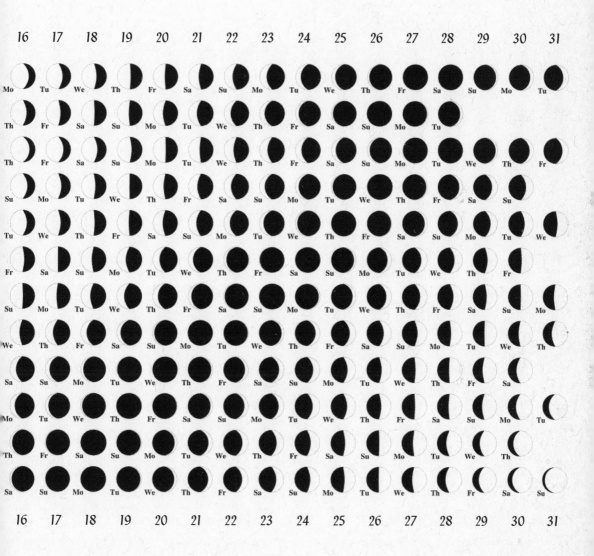

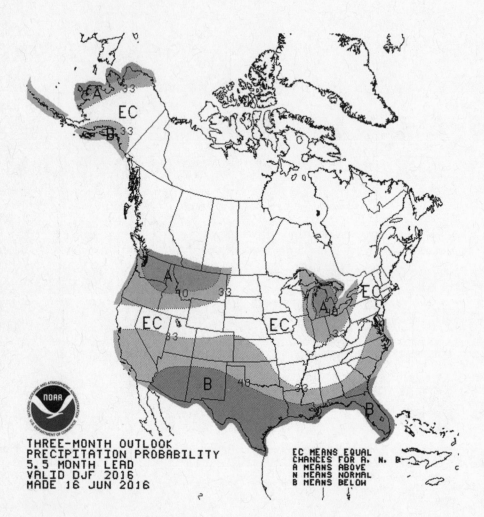

THREE-MONTH OUTLOOK
PRECIPITATION PROBABILITY
5.5 MONTH LEAD
VALID DJF 2016
MADE 16 JUN 2016

EC MEANS EQUAL
CHANCES FOR A, N, B
A MEANS ABOVE
N MEANS NORMAL
B MEANS BELOW

NOAA, 2016

PREDICTED PRECIPITATION DECEMBER-FEBRUARY

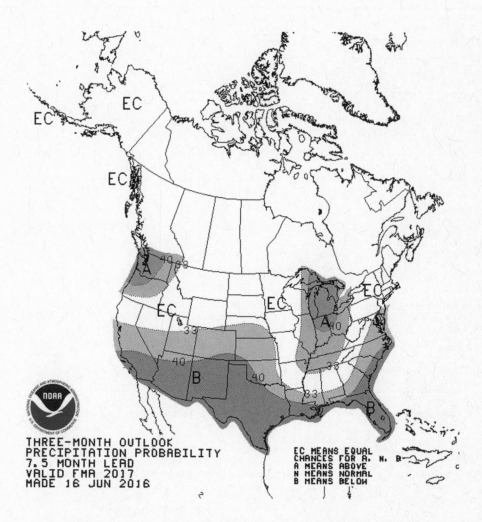

NOAA, 2016

THREE-MONTH OUTLOOK
PRECIPITATION PROBABILITY
7.5 MONTH LEAD
VALID FMA 2017
MADE 16 JUN 2016

EC MEANS EQUAL
CHANCES FOR A. N. B.
A MEANS ABOVE
N MEANS NORMAL
B MEANS BELOW

PREDICTED PRECIPITATION MARCH–MAY

THE HEAVENS

DATA EXCERPTED FROM THE U.S. NAVAL OBSERVATORY AND SEA AND SKY'S ASTRONOMY REFERENCE GUIDE

A NOTE ON TIME The astronomical data are expressed in the scale of universal time (UT); this is also known as Greenwich mean time (GMT) and is the standard time of the Greenwich meridian (0° of longitude). A time in UT may be converted to local mean time by the addition of east longitude (or subtraction of west longitude), where the longitude of the place is expressed in time-measure at the rate of one hour for every 15°.

ASTRONOMICAL TERMS

PERIHELION: the instant when a planet is closest to the Sun.

APHELION: the instant when a planet is farthest from the Sun.

PERIGEE: the instant when the Moon is closest to Earth.

APOGEE: the instant when the Moon is farthest from Earth.

INFERIOR CONJUNCTION: the instant when a planet passes between Earth and the Sun (Mercury or Venus).

SUPERIOR CONJUNCTION: the instant when a planet passes on the opposite side of the Sun from Earth (Mercury or Venus).

GREATEST ELONGATION: elongation is the apparent angle between a planet and the Sun as seen from Earth; during eastern elongation (E), the planet appears as an evening star; during western elongation (W), the planet appears as a morning star.

OPPOSITION: the instant when a planet appears opposite the Sun as seen from Earth.

CONJUNCTION: the instant when a planet appears closest the Sun as seen from Earth.

OCCULTATION: the Moon occults or eclipses a star or planet.

ASCENDING NODE: the point where a planet crosses from the southern to northern portion of its orbit.

DESCENDING NODE: the point where a planet crosses from the northern to the southern portion.

PRINCIPAL PHENOMENA OF SUN AND MOON

THE SUN

	d h		d h m		d h m
Perigee ... Jan. 4 14		Equinoxes ... Mar. 20 10 29		.... Sept. 22 20 02	
Apogee ... July 3 20		Solstices .... June 21 04 24		.... Dec. 21 16 28	

ECLIPSES

A penumbral eclipse of the Moon: Feb. 10–11; Western Asia, Africa, Europe, Greenland, South America, North America, and parts of the Pacific Ocean. **An annular eclipse of the Sun**: Feb. 26; SE Pacific Ocean, S. half of S. America, most of Antarctica, Africa (except northern parts). **A partial eclipse of the Moon**: Aug. 7; Western Pacific Ocean, Oceania, Australasia, Asia, Africa, Europe, easternmost tip of South America. **A total eclipse of the Sun**: Aug. 21; Hawaii, NE Pacific Ocean, North America, Central America, northern parts of South America, westernmost tip of Europe and W. Africa.

LUNAR PHENOMENA

d h	d h	d h		d h	d h	d h
Jan. 10 06	May 26 01	Oct. 9 06		Jan. 22 00	June 8 22	Oct. 25 02
Feb. 6 14	June 23 11	Nov. 6 00		Feb. 18 21	July 6 04	Nov. 21 19
Mar. 3 08	July 21 17	Dec. 4 09		Mar. 18 17	Aug. 2 18	Dec. 19 01
Mar. 30 13	Aug. 18 13			Apr. 15 10	Aug. 30 11	
Apr. 27 16	Sept. 13 16			May 12 20	Sept. 27 07	

PLANETARY PHENOMENA

MERCURY

	d h	d h	d h	d h
Stationary	Jan. 8 10	May 2 14	Sept. 4 16	Dec. 23 03
Greatest elongation West	Jan. 19 10 (24°)	May 17 23 (26°)	Sept. 12 10 (18°)	—
Superior conjunction ...	Mar. 7 00	June 21 14	Oct. 8 21	—
Greatest elongation East	Apr. 1 10 (19°)	July 30 05 (27°)	Nov. 24 00 (22°)	—
Stationary	Apr. 10 01	Aug. 12 06	Dec. 3 08	—
Inferior conjunction ...	Apr. 20 06	Aug. 26 21	Dec. 13 02	—

VENUS

		d h			d h
Greatest elongation East	Jan.	12 13 (47°)	Stationary	Apr.	13 00
Greatest illuminated extent	Feb.	17 07	Greatest illuminated extent	Apr.	30 04
Stationary	Mar.	2 14	Greatest elongation West	June	3 13 (46°)
Inferior conjunction ...	Mar.	25 10			

EARTH

	d h		d h m		d h m
Perihelion ...	Jan. 4 14	Equinoxes ...	Mar. 20 10 29 ...	...	Sept. 22 20 02
Aphelion ...	July 3 20	Solstices ...	June 21 04 24 ...	...	Dec. 21 16 28

SUPERIOR PLANETS

	Conjunction	Stationary	Opposition	Stationary
	d h	d h	d h	d h
Mars	July 27 01	—	—	—
Jupiter	Oct. 26 18	Feb. 6 19	Apr. 7 22	June 10 05
Saturn	Dec. 21 21	Apr. 6 05	June 15 10	Aug. 25 15
Uranus	Apr. 14 06	Aug. 3 10	Oct. 19 18	—
Neptune	Mar. 2 03	June 16 23	Sept. 5 05	Nov. 22 21

The vertical bars indicate where the dates for the planet are not in chronological order.

HELIOCENTRIC PHENOMENA

	Aphelion	Perihelion	Descending Node	Greatest Lat. South	Ascending Node	Greatest Lat. North
Mercury	Feb. 7	Mar. 23	Jan. 28	Feb. 27	Mar. 18	Jan. 4
	May 6	June 19	Apr. 26	May 26	June 14	Apr. 2
	Aug. 2	Sept. 15	July 23	Aug. 22	Sept. 10	June 29
	Oct. 29	Dec. 12	Oct. 19	Nov. 18	Dec. 7	Sept. 25
	—	—	—	—	—	Dec. 22
Venus	—	Feb. 20	May 9	July 5	Jan. 17	Mar. 14
	June 12	Oct. 3	Dec. 19	—	Aug. 30	Oct. 24
Mars	Oct. 7	—	—	—	Feb. 27	Aug. 30

Jupiter: Aphelion, Feb. 17
Saturn, Uranus, Neptune: None in 2017

VISIBILITY OF PLANETS

MERCURY can only be seen low in the east before sunrise, or low in the west after sunset (about the time of beginning or end of civil twilight). It is visible in the mornings between the following approximate dates: January 4 to February 24, April 29 to June 14, September 4 to September 28, and December 19 to December 31. The planet is brighter at the end of each period, (the best conditions in northern latitudes occur in mid-September and in late December, and in southern latitudes in the second half of May). It is visible in the evenings between the following approximate dates: March 16 to April 12, June 29 to August 20, and October 23 to December 7. The planet is brighter at the beginning of each period (the best conditions in northern latitudes occur from late March to early April, and in southern latitudes from mid-July to mid-August).

VENUS is a brilliant object in the evening sky until in the second half of March when it becomes too close to the Sun for observation. It reappears in late March as a morning star and can be seen in the morning sky until late November when it again becomes too close to the Sun for observation. Venus is in conjunction with Mars on October 5 and with Jupiter on November 13.

MARS can be seen only in the evening sky until early June passing through Aquarius, Pisces from late January, into Aries in early March, Taurus in mid-April (passing 6° N of Aldebaran on May 7), and into Gemini in early June. From the start of the second week of June it becomes too close to the Sun for observation and reappears in the morning sky in mid-September in Leo, moves into Virgo in mid-October (passing 3° N of Spica on November 28) and then into Libra in late December. Mars is in conjunction with Mercury on September 16 and with Venus on October 5. The reddish tint of Mars should assist in its identification.

JUPITER can be seen in Virgo from the beginning of the year and from mid-January can be seen for more than half the night (passing 4° N of Spica on January 20 and again 4° N of Spica on February 23). It is at opposition on April 7 when it can be seen throughout the night. From early July it can only be seen in the evening sky (passing 3° N of Spica on September 5) and from mid-October it becomes too close to the Sun for observation. It reappears in the morning sky in the second week of November and passes into Libra in mid-November. Jupiter is in conjunction with Venus on November 13.

SATURN rises shortly before sunrise at the beginning of the year in Ophiucus, passing into Sagittarius in late February and can only be seen in the morning sky until mid-March. Its westward elongation gradually increases, passing into Ophiucus again in the second half of May, and is at opposition on June 15, when it can be seen throughout the night. Its eastward elongation gradually decreases, and from mid-September until early December it can only be seen in the evening sky. It returns into

Sagittarius in mid-November and in early December it becomes too close to the Sun for observation for the remainder of the year. Saturn is in conjunction with Mercury on November 28.

URANUS is visible at the beginning of the year in Pisces and remains in this constellation throughout the year. From mid-January it can only be seen in the evening sky until late March when it becomes too close to the Sun for observation. It reappears in early May in the morning sky and is at opposition on Oct. 19. Its eastward elongation gradually decreases, and Uranus can be seen for more than half the night.

NEPTUNE is visible at the beginning of the year in the evening sky in Aquarius and remains in this constellation throughout the year. In the second week of February it becomes too close to the Sun for observation and reappears in the second half of March in the morning sky. Neptune is at opposition on September 5 and from early December can only be seen in the evening sky.

DO NOT CONFUSE (1) Mercury with Mars in mid-September and with Saturn in late November to early December; on both occasions Mercury is the brighter object. (2) Venus with Mars in late September to mid-October and with Jupiter in mid-November; on both occasions Venus is the brighter object. (3) Mars with Jupiter in late December when Jupiter is the brighter object.

VISIBILITY OF PLANETS IN MORNING AND EVENING TWILIGHT

	MORNING	EVENING
VENUS	March 30–November 28	January 1–March 22
MARS	September 12–December 31	January 1–June 7
JUPITER	January 1 – April 7 November 9–December 31	April 7–October 13
SATURN	January 1–June 15	June 15–December 5

2017 CALENDAR OF ASTRONOMICAL EVENTS

Full Moon. Photographed by Prof. Rutherford.
Copyrighted, 1897, by T. W. Ingersoll.

January 3, 4: Quadrantids Meteor Shower. The Quadrantids is an above average shower, with up to forty meteors per hour at its peak. It is thought to be produced by dust grains left behind by an extinct comet known as 2003 EH1, which was discovered in 2003. The shower runs annually from January 1–5. It peaks this year on the night of the 3rd and morning of the 4th. The first quarter moon will set shortly after midnight leaving fairly dark skies for what could be a good show. Best viewing will be from a dark location after midnight. Meteors will radiate from the constellation Bootes but can appear anywhere in the sky.

January 12: Full Moon. The Moon will be located on the opposite side of the Earth as the Sun and its face will be will be fully illuminated. This phase occurs at 11:34 UTC. This full moon was known by early Native Americans as the Full Wolf Moonbecause this was the time of year when hungry wolf packs howled outside their camps. This moon has also been know as the Old Moon and the Moon After Yule.

January 12: Venus at Greatest Eastern Elongation. The planet Venus reaches greatest eastern elongation of 47.1 degrees from the Sun. This is the best time to view Venus since it will be at its highest point above the horizon in the evening sky. Look for the bright planet in the western sky after sunset.

January 19: Mercury at Greatest Western Elongation.

49

The planet Mercury reaches greatest western elongation of 24.1 degrees from the Sun. This is the best time to view Mercury since it will be at its highest point above the horizon in the morning sky. Look for the planet low in the eastern sky just before sunrise.

January 28: New Moon. The Moon will located on the same side of the Earth as the Sun and will not be visible in the night sky. This phase occurs at 00:07 UTC. This is the best time of the month to observe faint objects such as galaxies and star clusters because there is no moonlight to interfere.

February 11: Full Moon. The Moon will be located on the opposite side of the Earth as the Sun and its face will be will be fully illuminated. This phase occurs at 00:33 UTC. This full moon was known by early Native Americans as the Full Snow Moon because the heaviest snows usually fell during this time of the year. Since hunting is difficult, this moon has also been known by some tribes as the Full Hunger Moon, as the harsh weather made hunting difficult.

February 11: Penumbral Lunar Eclipse. A penumbral lunar eclipse occurs when the Moon passes through the Earth's partial shadow, or penumbra. During this type of eclipse the Moon will darken slightly but not completely. The eclipse will be visible throughout most of eastern South America, eastern Canada, the Atlantic Ocean, Europe, Africa, and western Asia. (NASA Map and Eclipse Information).

February 26: New Moon. The Moon will located on the same side of the Earth as the Sun and will not be visible in the night sky. This phase occurs at 14:59 UTC. This is the best time of the month to observe faint objects such as galaxies and star clusters because there is no moonlight to interfere.

February 26: Annular Solar Eclipse. An annular solar eclipse occurs when the Moon is too far away from the Earth to completely cover the Sun. This results in a ring of light around the darkened Moon. The Sun's corona is not visible during an annular eclipse. The path of the eclipse will begin off the coast of Chile and pass through southern Chile and southern Argentina, across the southern Atlantic Ocean, and into Angola and Congo in Africa. A partial eclipse will be visible throughout parts of southern South America and southwestern Africa. (NASA Map and Eclipse Information; NASA Interactive Google Map).

March 12: Full Moon. The Moon will be located on the opposite side of the Earth as the Sun and its face will be will be fully illuminated. This phase occurs at 14:54 UTC. This full moon was known by early Native Americans as the Full Worm Moon because this was the time of year when the ground would begin to soften and the earthworms would reappear. This moon has also been known as the Full Crow Moon, the Full Crust Moon, the Full Sap Moon, and the Lenten Moon.

March 20: March Equinox. The March equinox occurs at 10:29 UTC. The Sun will shine directly on the equator and there will be nearly equal amounts of day and night throughout the world. This is also the first day of spring (vernal equinox) in the Northern Hemisphere and the first day of fall (autumnal equinox) in the Southern Hemisphere.

March 28: New Moon. The Moon will located on the same side of the Earth as the Sun and will not be visible in the night sky. This phase occurs at 02:58 UTC. This is the best time of the month to observe faint objects such as galaxies and star clusters because there is no moonlight to interfere.

April 1: Mercury at Greatest Eastern Elongation. The planet Mercury reaches greatest eastern elongation of nineteen degrees from the Sun. This is the best time to view Mercury since it will be at its highest point above the horizon in the evening sky. Look for the planet low in the western sky just after sunset.

April 7: Jupiter at Opposition. The giant planet will be at its closest approach to Earth and its face will be fully illuminated by the Sun. It will be brighter than any other time of the year and will be visible all night long. This is the best time to view and photograph Jupiter and its moons. A medium-sized telescope should be able to show you some of the details in Jupiter's cloud bands. A good pair of binoculars should allow you to see Jupiter's four largest moons, appearing as bright dots on either side of the planet.

April 11: Full Moon. The Moon will be located on the opposite side of the Earth as the Sun and its face will be will be fully illuminated. This phase occurs at 06:08 UTC. This full moon was known by early Native Americans as the Full Pink Moon because it marked the appearance of the moss pink, or wild ground phlox, which is one of the first spring flowers. This moon has also been known as the Sprouting Grass Moon, the Growing Moon, and the Egg Moon. Many coastal peoples called it the Full Fish Moon because this was the time that the shad swam upstream to spawn.

April 22, 23: Lyrids Meteor Shower. The Lyrids is an average shower, usually producing about twenty meteors per hour at its peak. It is produced by dust particles left behind by comet C/1861 G1 Thatcher, which was discovered in 1861. The shower runs annually from April 16–25. It peaks this year on the night of the night of the 22nd

and morning of the 23rd. These meteors can sometimes produce bright dust trails that last for several seconds. The crescent moon should not be too much of a problem this year. Skies should still be dark enough for a good show. Best viewing will be from a dark location after midnight. Meteors will radiate from the constellation Lyra but can appear anywhere in the sky.

April 26: New Moon. The Moon will located on the same side of the Earth as the Sun and will not be visible in the night sky. This phase occurs at 12:17 UTC. This is the best time of the month to observe faint objects such as galaxies and star clusters because there is no moonlight to interfere.

May 6, 7: Eta Aquarids Meteor Shower. The Eta Aquarids is an above average shower, capable of producing up to sixty meteors per hour at its peak. Most of the activity is seen in the Southern Hemisphere. In the Northern Hemisphere, the rate can reach about thirty meteors per hour. It is produced by dust particles left behind by comet Halley, which has known and observed since ancient times. The shower runs annually from April 19 to May 28. It peaks this year on the night of May 6 and the morning of the May 7. The waxing gibbous moon will block out many of the fainter meteors this year. But if you are patient, you should be able to catch quite a few of the brighter ones. Best viewing will be from a dark location after midnight. Meteors will radiate from the constellation Aquarius but can appear anywhere in the sky.

May 10: Full Moon. The Moon will be located on the opposite side of the Earth as the Sun and its face will be will be fully illuminated. This phase occurs at 21:42 UTC. This full moon was known by early Native Americans as the Full Flower Moon because this was the time of year when spring flowers appeared in abundance. This moon has also been known as the Full Corn Planting Moon and the Milk Moon.

May 17: Mercury at Greatest Western Elongation. The planet Mercury reaches greatest western elongation of 25.8 degrees from the Sun. This is the best time to view Mercury since it will be at its highest point above the horizon in the morning sky. Look for the planet low in the eastern sky just before sunrise.

May 25: New Moon. The Moon will located on the same side of the Earth as the Sun and will not be visible in the night sky. This phase occurs at 19:45 UTC. This is the best time of the month to observe faint objects such as galaxies and star clusters because there is no moonlight to interfere.

June 3: Venus at Greatest Western Elongation. The planet Venus reaches greatest eastern elongation of 45.9 degrees from the Sun. This is the best time to view Venus since it will be at its highest point above the horizon in the morning sky. Look for the bright planet in the eastern sky before sunrise.

June 9: Full Moon. The Moon will be located on the opposite side of the Earth as the Sun and its face will be will be fully illuminated. This phase occurs at 13:10 UTC. This full moon was known by early Native Americans as the Full Strawberry Moon because it signaled the time of year to gather ripening fruit. It also coincides with the peak of the strawberry harvesting season. This moon has also been known as the Full Rose Moon and the Full Honey Moon.

June 15: Saturn at Opposition. The ringed planet will be at its closest approach to Earth and its face will be fully illuminated by the Sun. It will

be brighter than any other time of the year and will be visible all night long. This is the best time to view and photograph Saturn and its moons. A medium-sized or larger telescope will allow you to see Saturn's rings and a few of its brightest moons.

June 21: June Solstice. The June solstice occurs at 04:24 UTC. The North Pole of the earth will be tilted toward the Sun, which will have reached its northernmost position in the sky and will be directly over the Tropic of Cancer at 23.44 degrees north latitude. This is the first day of summer (summer solstice) in the Northern Hemisphere and the first day of winter (winter solstice) in the Southern Hemisphere.

June 24: New Moon. The Moon will located on the same side of the Earth as the Sun and will not be visible in the night sky. This phase occurs at 02:31 UTC. This is the best time of the month to observe faint objects such as galaxies and star clusters because there is no moonlight to interfere.

July 9: Full Moon. The Moon will be located on the opposite side of the Earth as the Sun and its face will be will be fully illuminated. This phase occurs at 04:07 UTC. This full moon was known by early Native Americans as the Full Buck Moon because the male buck deer would begin to grow their new antlers at this time of year. This moon has also been known as the Full Thunder Moon and the Full Hay Moon.

July 23: New Moon. The Moon will located on the same side of the Earth as the Sun and will not be visible in the night sky. This phase occurs at 09:46 UTC. This is the best time of the month to observe faint objects such as galaxies and star clusters because there is no moonlight to interfere.
July 28, 29: Delta Aquarids Meteor Shower.

The Delta Aquarids is an average shower that can produce up to twenty meteors per hour at its peak. It is produced by debris left behind by comets Marsden and Kracht. The shower runs annually from July 12 to August 23. It peaks this year on the night of July 28 and morning of July 29. The crescent moon will set by midnight, leaving dark skies for what should be a good early morning show. Best viewing will be from a dark location after midnight. Meteors will radiate from the constellation Aquarius but can appear anywhere in the sky.

July 30: Mercury at Greatest Eastern Elongation. The planet Mercury reaches greatest eastern elongation of 27.2 degrees from the Sun. This is the best time to view Mercury since it will be at its highest point above the horizon in the evening sky. Look for the planet low in the western sky just after sunset.

August 7: Full Moon. The Moon will be located on the opposite side of the Earth as the Sun and its face will be will be fully illuminated. This phase occurs at 18:11 UTC. This full moon was known by early Native Americans as the Full Sturgeon Moon because the large sturgeon fish of the Great Lakes and other major lakes were more easily caught at this time of year. This moon has also been known as the Green Corn Moon and the Grain Moon.

August 7: Partial Lunar Eclipse. A partial lunar eclipse occurs when the Moon passes through the Earth's partial shadow, or penumbra, and only a portion of it passes through the darkest shadow, or umbra. During this type of eclipse a part of the Moon will darken as it moves through the Earth's shadow. The eclipse will be visible throughout most of eastern Africa, central Asia, the Indian Ocean, and Australia. (NASA Map and Eclipse Information).

August 12, 13: Perseids Meteor Shower. The Perseids is one of the best meteor showers to observe, producing up to sixty meteors per hour at its peak. It is produced by comet Swift-Tuttle, which was discovered in 1862. The Perseids are famous for producing a large number of bright meteors. The shower runs annually from July 17 to August 24. It peaks this year on the night of August 12 and the morning of August 13. The waning gibbous moon will block out many of the fainter meteors this year, but the Perseids are so bright and numerous that it should still be a good show. Best viewing will be from a dark location after midnight. Meteors will radiate from the constellation Perseus but can appear anywhere in the sky.

August 21: New Moon. The Moon will located on the same side of the Earth as the Sun and will not be visible in the night sky. This phase occurs at 18:30 UTC. This is the best time of the month to observe faint objects such as galaxies and star clusters because there is no moonlight to interfere.

August 21: Total Solar Eclipse. A total solar eclipse occurs when the moon completely blocks the Sun, revealing the Sun's beautiful outer atmosphere known as the corona. This is a rare, once-in-a-lifetime event for viewers in the United States. The last total solar eclipse visible in the continental United States occurred in 1979 and the next one will not take place until 2024. The path of totality will begin in the Pacific Ocean and travel through the center of the United States. The total eclipse will be visible in parts of Oregon, Idaho, Wyoming, Nebraska, Missouri, Kentucky, Tennessee, North Carolina, and South Carolina before ending in the Atlantic Ocean. A partial eclipse will be visible in most of North America and parts of northern South America. (NASA Map and Eclipse Information: Detailed Zoomable Map of Eclipse Path).

September 5: Neptune at Opposition. The blue giant planet will be at its closest approach to Earth and its face will be fully illuminated by the Sun. It will be brighter than any other time of the year and will be visible all night long. This is the best time to view and photograph Neptune. Due to its extreme distance from Earth, it will only appear as a tiny blue dot in all but the most powerful telescopes.

September 6: Full Moon. The Moon will be located on the opposite side of the Earth as the Sun and its face will be will be fully illuminated. This phase occurs at 07:03 UTC. This full moon was known by early Native Americans as the Full Corn Moon because the corn is harvested around this time of year.

September 12: Mercury at Greatest Western Elongation. The planet Mercury reaches greatest western elongation of 17.9 degrees from the Sun. This is the best time to view Mercury since it will be at its highest point above the horizon in the morning sky. Look for the planet low in the eastern sky just before sunrise.

September 20: New Moon. The Moon will located on the same side of the Earth as the Sun and will not be visible in the night sky. This phase occurs at 05:30 UTC. This is the best time of the month to observe faint objects such as galaxies and star clusters because there is no moonlight to interfere.

September 22: September Equinox. The September equinox occurs at 20:02 UTC. The Sun will shine directly on the equator and there will be nearly equal amounts of day and night throughout the world. This is also the first day of fall (autumnal equinox) in the Northern Hemisphere and the first day of spring (vernal equinox) in the Southern Hemisphere.

October 5: Full Moon. Moon will be directly opposite the Earth from the Sun and will be fully illuminated as seen from Earth. This phase occurs at 18:40 UTC. This full moon was known by early Native American tribes as the Full Hunters Moon because at this time of year the leaves are falling and the game is fat and ready to hunt. This moon has also been known as the Travel Moon and the Blood Moon. This moon is also known as the Harvest Moon. The Harvest Moon is the full moon that occurs closest to the September equinox each year.

October 7: Draconids Meteor Shower. The Draconids is a minor meteor shower producing only about ten meteors per hour. It is produced by dust grains left behind by comet 21P Giacobini-Zinner, which was first discovered in 1900. The Draconids is an unusual shower in that the best viewing is in the early evening instead of early morning like most other showers. The shower runs annually from October 6–10 and peaks this year on the the night of the 7th. Unfortunately, the nearly full moon will block all but the brightest meteors this year. If you are extremely patient, you may be able to catch a few good ones. Best viewing will be in the early evening from a dark location far away from city lights. Meteors will radiate from the constellation Draco but can appear anywhere in the sky.

October 19: New Moon. The Moon will located on the same side of the Earth as the Sun and will not be visible in the night sky. This phase occurs at 19:12 UTC. This is the best time of the month to observe faint objects such as galaxies and star clusters because there is no moonlight to interfere.

October 19: Uranus at Opposition. The blue-green planet will be at its closest approach to Earth and its face will be fully illuminated by the

Sun. It will be brighter than any other time of the year and will be visible all night long. This is the best time to view Uranus. Due to its distance, it will only appear as a tiny blue-green dot in all but the most powerful telescopes.

October 21, 22: Orionids Meteor Shower. The Orionids is an average shower producing up to twenty meteors per hour at its peak. It is produced by dust grains left behind by comet Halley, which has been known and observed since ancient times. The shower runs annually from October 2 to November 7. It peaks this year on the night of October 21 and the morning of October 22. The crescent moon will set early in the evening leaving dark skies for what should be a good show. Best viewing will be from a dark location after midnight. Meteors will radiate from the constellation Orion but can appear anywhere in the sky.

November 4: Full Moon. The Moon will be located on the opposite side of the Earth as the Sun and its face will be will be fully illuminated. This phase occurs at 05:23 UTC. This full moon was known by early Native Americans as the Full Beaver Moon because this was the time of year to set the beaver traps before the swamps and rivers froze. It has also been known as the Frosty Moon and the Hunter's Moon.

November 4, 5: Taurids Meteor Shower. The Taurids is a long-running minor meteor shower producing only about five to ten meteors per hour. It is unusual in that it consists of two separate streams. The first is produced by dust grains left behind by Asteroid 2004 TG10. The second stream is produced by debris left behind by Comet 2P Encke. The shower runs annually from September 7 to December 10. It peaks this year on the the night of November 4.

Unfortunately the glare from the full moon will block out all but the brightest meteors. If you are extremely patient, you may still be able to catch a few good ones. Best viewing will be just after midnight from a dark location far away from city lights. Meteors will radiate from the constellation Taurus but can appear anywhere in the sky.

November 13: Conjunction of Venus and Jupiter. A spectacular conjunction of Venus and Jupiter will be visible in the evening sky. The two bright planets will be extremely close, appearing only 0.3 degrees apart. Look for this impressive pairing in the Eastern sky just before sunrise.

November 17, 18: Leonids Meteor Shower. The Leonids is an average shower, producing up to fifteen meteors per hour at its peak. This shower is unique in that it has a cyclonic peak about every thirty-three years where hundreds of meteors per hour can be seen. That last of these occurred in 2001. The Leonids is produced by dust grains left behind by comet Tempel-Tuttle, which was discovered in 1865. The shower runs annually from November 6–30. It peaks this year on the night of the 17th and morning of the 18th. The nearly new moon will not be a problem this year. Skies should be dark enough for what should be good show. Best viewing will be from a dark location after midnight. Meteors will radiate from the constellation Leo but can appear anywhere in the sky.

November 18: New Moon. The Moon will located on the same side of the Earth as the Sun and will not be visible in the night sky. This phase occurs at 11:42 UTC. This is the best time of the month to observe faint objects such as galaxies and star clusters because there is no moonlight to interfere.

November 24: Mercury at Greatest Eastern Elongation. The planet Mercury reaches greatest eastern elongation of 22.0 degrees from the Sun. This is the best time to view Mercury since it will be at its highest point above the horizon in the evening sky. Look for the planet low in the western sky just after sunset.

December 3: Full Moon, Supermoon. The Moon will be located on the opposite side of the Earth as the Sun and its face will be will be fully illuminated. This phase occurs at 15:47 UTC. This full moon was known by early Native Americans as the Full Cold Moon because this is the time of year when the cold winter air settles in and the nights become long and dark. This moon has also been known as the Full Long Nights Moon and the Moon Before Yule. This is also the only supermoon for 2017. The Moon will be at its closest approach to the Earth and may look slightly larger and brighter than usual.

December 13, 14: Geminids Meteor Shower. The Geminids is the king of the meteor showers. It is considered by many to be the best shower in the heavens, producing up to 120 multicolored meteors per hour at its peak. It is produced by debris left behind by an asteroid known as 3200 Phaethon, which was discovered in 1982. The shower runs annually from December 7–17. It peaks this year on the night of the 13th and morning of the 14th. The waning crescent moon will be no match for the Geminids this year. The skies should still be dark enough for an excellent show. Best viewing will be from a dark location after midnight. Meteors will radiate from the constellation Gemini but can appear anywhere in the sky.

December 18: New Moon. The Moon will located on the same side of the Earth as the Sun and will not be visible in the night sky. This phase occurs at 06:30 UTC. This is the best time of the month to observe faint objects such as galaxies and star clusters because there is no moonlight to interfere.

December 21: December Solstice. The December solstice occurs at 16:28 UTC. The South Pole of the earth will be tilted toward the Sun, which will have reached its southernmost position in the sky and will be directly over the Tropic of Capricorn at 23.44 degrees south latitude. This is the first day of winter (winter solstice) in the Northern Hemisphere and the first day of summer (summer solstice) in the Southern Hemisphere.

December 21, 22: Ursids Meteor Shower. The Ursids is a minor meteor shower producing about five to ten meteors per hour. It is produced by dust grains left behind by comet Tuttle, which was first discovered in 1790. The shower runs annually from December 17–25. It peaks this year on the the night of the 21st and morning of the 22nd. The crescent moon will set early in the evening leaving dark skies for optimal observing. Best viewing will be just after midnight from a dark location far away from city lights. Meteors will radiate from the constellation Ursa Minor but can appear anywhere in the sky.

JANUARY

WATER

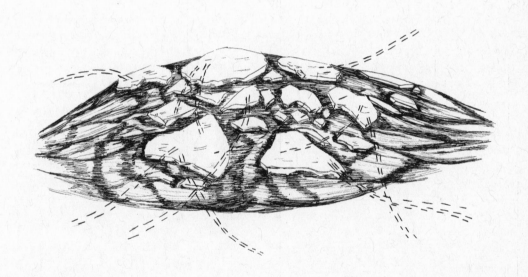

Jacob van Ruisdael

CONSCIOUSNESS IN COMMON

LORIG HAWKINS

To restore a more resilient foundation for society to coexist with all life, we must romanticize the small moments within the ecologies of our farms. We must reengage with our place and our lives and not just seek beautiful moments but presence in every moment. Presence, companionship, ritual, tradition, and authentic connection with our places, people, and moments in the field can restore mindful complexity to thinking about each other.

Here are some of the moments that changed the romance and dimensions of my thoughts and feelings about this common life.

COMPANIONSHIP AS COMMONS

Theo, my collie-heeler, hurriedly eats his breakfast while I slowly prepare my morning coffee. The ping of the spoon against the edge of the ceramic mug plays off the banging of the metal dog bowl against the wall. We finish simultaneously—he licks the remainder of the bowl and I take the first sip of the warm, bitter black coffee. He leads the way as we move toward the back door and I turn the cold metal doorknob. Theo, a black and white blur, brushes against my pant leg, rushes outside, chasing some

creature in the lawn as I step down the porch steps to sit and tie my boots. I stand up, ready to walk the fields, and Theo returns as if on cue, standing in front of me and waiting for the signal to start the ritual walk of the fields.

We head east toward the sun, squinting at the brightness and happy in the warmth. Like I do most mornings, I think about this ritual and its place in my day. We're in the commons, learning it, paying attention together.

PRESENT-MOMENT AWARENESS AS COMMONS

The property of the first farm I managed was beautiful, worn, and cluttered. Attention to detail to the buildings and storage areas relayed someone was tending to the space. My team and I would stand in the well-worn walk-aisles between rows of vegetables and look around the area, taking our gaze from the broad open space of the fields around us to the plants and bugs mere inches from our feet.

As we worked there for weeks, the three-acre space became the focal point of the world. Caught up in activity, as a farm manager, my vision would narrow. I could forget where I was as all the demands and tasks of the day held my attention and focus. I fell into a bad habit of thinking into the future, always trying to anticipate the next task and the next move. I wanted to create a seamless flow of bodies as crew members transitioned from task to task—not a moment wasted in idleness.

The birds would expand my vision again; bring me back to the whole. Their sharp, distant calls would pull me into the present moment, back to the solace that I sought from the place and the vocation. The rustle of wings and voices carrying through the wind would snap my thoughts open.

Maybe it is was that they all moved and swayed together, or the pressure and breeze in the air, or maybe just the simple act of looking up, but it simultaneously connected me to the whole while putting me in the immediate presence of my body there, in the fields, at that moment gazing up.

Now I always allow my mind to float to the space above. This purposeful ritual with the birds keeps my daily awareness in both the sky and the soil— the precious commons of complete presence.

GRATITUDE AS COMMONS

The above experiences could not exist without the farming community—farmers, growers, leaders, volunteers, readers, eaters, dogs, supporters, doers, believers, activists, grandmothers, cats, and the unknown (and hopefully not forgotten) rich field of possibility and potential that make up the commons.

Dedicated to Theo, whose body left this world July 2015, only to leave his spirit with us at the farm forever.

JOHN'S MURAL

BRIAN LAIDLAW

ascend the system
of ripples up

a watercolor which depicts
a dry deathless valley

the ladder rungs
ring wet and pulpy

you slip, you splatter
feel tongues of a glacier

like a hand, a face.
the mural

is watercolor on conifer,
watercolor on copperhead

and deadfall,
a valley submerged

in water, in color.
ascensionist, as you

approach the wall in a canvas
floating raft, careful:

your bones are
singing within you

and if you bequeath your body
to geology, you

are the world's to be pulped
and sheeted, cross-examined

and splayed. the pines have needles
unlike a man has hair,

the valley has pines
unlike a city has men,

and the stones have time
unlike a man has time.

which means there's more to the mural
than the currents of bloodshed,

but there's also the bloodshed.

From *The Imaginary Climbs of Hetch Hetchy.*

FIG. 38. *Eisenia bicyclis* (× 0.6) (After Okamura)

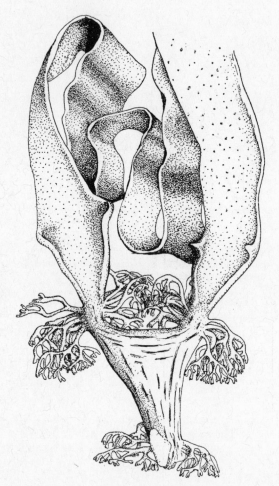

FIG. 33.   *Arthrothamnus bifidus* ($\times \frac{1}{2}$)
(After Okamura)

MONEY

JANE MEAD

Someone had the idea of getting more water
released beneath the Don Pedro Dam
into the once-green Tuolumne—

so the minnows could have some wiggle room,
so the salmon could lunge far enough up
to spawn, so that there would be more salmon

in the more water below the dam.
But it wasn't possible—by then the water
didn't belong to the salmon anymore, by then

the water didn't even belong to the river.
The water didn't belong to the water.

From Jane Mead, *Money Money Money Water Water Water*
(Farmington, ME: Alice James Books, 2014). Reprinted with permission.

Martha Shaw

Farm Preserve Note; Deli Dollar

Editor's note: These are examples of some alternative currencies. This one called the Deli Dollar predated the Berkshare and was invented to help capitalize the move of a beloved local deli down the street. The idea with local currencies is to cycle more wealth within the community, supporting all the businesses and people who are connected together in relations of exchange—rather than to participate in a global, liquid, anonymous economy that tends to siphon money offsite and into the control of behemoths such

Barclay's Bank Note; NASA satellite image of Amazon River mouth

as Bayer/Monsanto, AT&T/Time Warner, and Syngenta/Chem China. Some farms are experimenting with redeemable coupons issued at the head of the season and redeemable in fall. Learn more about money systems from David Boyle's *Funny Money* and David Graeber's *Debt: The First 5000 Years*, as well as Michael Rowbotham's *The Grip of Death*, a wonderful book about the evolution of the U.S. monetary system. What if each watershed had its own currency? —SvTF

COMMUNAL WATER SOVEREIGNTY:

Acequias in New Mexico

NEW MEXICO ACEQUIA ASSOCIATION

For the Indo-Hispano people of New Mexico, the acequia is our birth canal, our sacred place of emergence, existence, and sustainability in the arid desert we call home. According to esteemed local author and historian Juan Estevan Arellano, the word *acequia* derives from the Arabic word *assaqiya*, meaning "someone bearing water." New Mexico's acequias are a synthesis of cultural, agricultural, and legal traditions inherited from the Moors of North Africa via Spain and the indigenous cultures of the Americas. The specific institution of the acequia took root in arid areas throughout present-day northern Mexico and the Southwestern United States from the sixteenth through the nineteenth centuries. An acequia is the physical structure for irrigation made from a common diversion from a spring or stream. These humble earthen ditches, constructed and dug with hand tools, distribute water through the simple force of gravity. One basic principle that acequias and similar systems have in common is that water is so essential to life that all living beings have a right to water for survival. Although specific customs and traditions vary by region, watershed, or village, the principle of sharing water underlies the basic operation of the system.

Water, like the common lands, historically has been viewed as a communal resource. Although a family might have a *derecho*, or a use-right, attached to their respective property, the water itself is treated as a mutual blessing. Use of water from the acequia is conditional on the obligation to contribute to the cooperative maintenance of the system, such as providing needed labor for cleaning and emergency repairs. Due to the chronic water scarcity facing communities, over time the acequias developed intricate customs for coping with water shortages that were localized. These customs and traditions are collectively referred to as the *repartimiento de agua*, or the dividing and sharing of the water. According to custom, water in abundance is shared equally. Likewise, the shortage of water is also shared, such that the amount of water allocated to landowners is reduced so that everyone gets at least some water. The delicate balance of rights and responsibilities in the process of sharing scarce water are characteristics of acequias that have persisted since their establishment.

The fundamental tenets of acequia management have changed little over the course of more than three centuries in New Mexico. As New Mexico became a territory of the United States, the acequia communal form of governance was codified in the state constitution, and acequias were recognized as political subdivisions of the state. This gives acequias the same authority as local governments, and thus the same responsibilities. Every acequia is governed by a commission of three members and one elected official, the *mayordomo*, who manages the day-to-day water distribution and maintenance. Today's acequias in New Mexico are concentrated predominantly in rural counties that have the highest rates of poverty and Indo-Hispano population. Up until two generations ago, many of our families made most of their livelihood from their *ranchitos*, or small scale farms and ranches. However, with the advent of a global food system that favors agribusiness, our communities adapted by becoming wage earners while also continuing their agricultural traditions on a part-time basis. Today, our communities face commodification of water and land, gentrification, and the challenges of keeping land-based livelihoods economically viable for low-income families.

While acequias have a long history of dealing with water scarcity, there is no precedent for the current water crisis that faces New Mexico. Decades of above-average rainfall coupled with an overdependence of groundwater for supplying cities and towns have set the stage for major reallocations of agricultural water. The cultural and legal tradition we inherited from Spain and Mexico recognized communal property. When we were confronted with Manifest Destiny,

Sharon Stewart

Editor's note: The spring cleaning of the acequia ditch is coordinated by the *mayordomo*. The *mayordomo* arranges the day, and invites all the *parciantes* to come to dig and clean and cut and smoothen the ditch and its banks. The *mayordomo* measures out with a long stick the assigned lengths (about six feet) that the shoveler must clean. He supervises to make sure it is cleaned to his liking. See more photographs online (Sharon Stewart, *Acequia Series*). —SvTF

communal land and water were privatized. Although our cultural worldview is that water is life and water is a *don divino*, or divine gift, New Mexico law treats water like a marketable commodity.

In New Mexico, all water is "appropriated," meaning that any new use of water can only come about at the expense of an existing use through the process known as a "water transfer." These water transfers have to be approved by the State Engineer, who is the governor-appointed water czar of New Mexico. Acequias have legally challenged water transfers through protests, and in 2003, through the efforts of political organizing by the New Mexico Acequia Association, a law was passed that gave acequias the legal authority to deny water transfers. Acequias have argued that treating water like a commodity would unravel our cultural heritage and our future water security.

According to a report written by the New Mexico Acequia Association, the prevailing assumption is that water for new growth will come from agricultural uses, thereby fueling an emerging "water market." According to studies of future supply and demand, acequia communities are projected to lose 30 to 60 percent of their water rights base and farmland to development in the next forty years. The concern is that acequias in areas with high water demands may be driven to extinction by water transfers. We are in the increasingly common position of having to defend our land and water base while also trying to rebuild our local food systems and reclaim our agricultural sovereignty. Protecting acequia water is a social justice issue where thousands of farmers, ranchers, and organizations are fighting tirelessly for the continuation of these ancient systems and ways of life.

The New Mexico Acequia Association was founded in 1988 on the belief that a strong statewide organization was necessary to enable us to secure our water rights and to develop the capabilities that our community-based acequia associations will need to contend with the political, legal, and economic challenges we face now and in the next century. We believe that our ability to grow our own food with the water from our acequias, the lands of our families, and the seeds of our ancestors makes us a free people. Our self-determination depends on retaining our ancestral lands and water under the stewardship that we inherited from our parents and grandparents. Our communities have been organizing, cultivating acequia lands with ancestral crops, and continuously improving farm and ranch soils to enhance the efficient use of water. Members and acequia families have been actively participating in the maintenance and governance of acequias and resisting transfers. *Acequieros* are constantly planting, learning, and adapting to the environment and stringent water laws. We have become like the giant cottonwoods living off the banks of the ditches, preserving the water we have in order to cultivate clusters of healthy *parciantes* who blossom and are too rooted to simply displace with a corporate hand.

United we dug the acequias, and united we defend *nuestra agua*!

Editor's note: The law of thirst—based in the Quran—means that water cannot be sold; it must be shared. This is embedded in the acequia legal governance structure, in which the elected officers hire the manager, the *mayordomo*, who is paid to take care of the ditch and stand guardian over its use and health. But he is not the boss; he is accountable to the elected board. Nobody owns the acequia. —SvTF

HOLDING WATER

LEAH SIENKOWSKI

R. de Salis

There was that time in the middle of July, my first season on a farm, during what happened to be the longest drought since the Dust Bowl. I was in southern Michigan—pulling weeds, killing potato beetles, but mostly moving hoses and sprinklers and transferring water from five-gallon jugs into watering cans to water the squash and tomato fields by hand. Except, of course, on that rare afternoon in July when it did begin to rain, and we stopped watering and instead ran around in the mud rescuing the garlic, which was curing outdoors, and the radio, which was sinking into the mud.

Then, again, last December, as I lay in

the bed of the dump-trailer, this time caked in mud myself and nested in piles of irrigation drip tape. Heather was driving the truck, and Emily and I were lying in the trailer wearing insulated Carhartts, winter coats, and several pairs of gloves. I distinctly remember those sweet expanses of time in which we were not dragging cold and muddy plastic out of the fields and lifting them over our heads, into the trailer, but were instead lying on top of them.

And again this March. I was following a herd of dairy goats into the high desert of Pie Town, New Mexico, when, all at once, they quickened their pace, merged into the arroyo as if it were a highway, and ran in a pack toward Cow Lake. Ears flapping and tails in the air, they reached the water, lined up along the edge, and drank.

In these moments I've felt connected to the drama and glory of life, specifically of water—a resource that has and does spark many complicated emotions but is not often discussed in depth or appreciated in full here in the Midwest, though we're surrounded by its fresh bodies and frozen icebergs.

In Pie Town, New Mexico, the average rainfall amounts to eleven inches each year, and we collected as much as we could—gathered by way of rooftops—and stored it for use throughout the year in large plastic vats. Here in the soggy Midwest, where the rainfall is close to the national average of thirty-seven inches, we direct it to the streets and into the sewers as if it were trash, while watering our lawns with sprinkler systems. In Pie Town, when it rained, the dirt road turned into a river, and the brittle desert soil became mud. The goats hated the

rain, and it could trap us at the dairy for days at a time, yet we treasured it—for drinking, washing, and cooking.

Like many resources, we are dispassionate and deluded about our reliance on water, and will be, until our land is brittle and our mouths parched. Someday though, like the horned desert lizard, we'll have to learn to arch our backs, welcome the storm, and reroute the shimmering drops into our mouths.

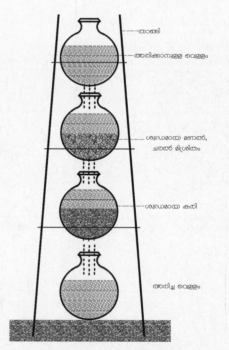

വെള്ളം ശുദ്ധീകരിക്കുന്നതിനുള്ള
മുച്ചട്ടിഅരിഷയുടെ മാതൃക

Water catchment

3-D OCEAN FARMING

BREN SMITH

Imagine an underwater garden: a mix of seaweeds and shellfish grown vertically using the entire water column, floating below the surface with low aesthetic impact, at once both using and preserving the ocean commons. Now picture this underwater garden mitigating climate change through carbon and nitrogen sequestration, and creating jobs for unemployed fishers who were victims of the crash of the commercial fishing market. Envision this underwater vertical garden having the capacity to grow thirty tons of sea vegetables and 250,000 shellfish per acre per year with no inputs, making it one of the most sustainable sources on the planet for food, animal feed, and biofuel.

After fifteen years of experimenting, I developed a model for this underwater garden, called 3-D Ocean Farming, designed to transform fishers into restorative ocean farmers in an era of climate change, overfishing, and food and job insecurity. It sprang from necessity, as I was a former fisherman in need of a new livelihood that could no longer come from overfished waters. After dropping out of school at age fourteen and working on fishing boats in the height of commercial fishing for nearly twenty years, catching McDonald's cod for many of them, I

was launched on a search for sustainability when the cod stocks crashed in the 1990s, my livelihood crashing along with them. I began my search on aquaculture farms, where we grew neither fish nor food and essentially ran concentrated animal feeding operations (CAFOs) at sea, but I realized that this strategy of monoculture, of vertical integration, and of growing around an industrial palate was unsustainable. Rather than asking what consumers want us to grow, our model asks what our oceans, our commons, are able to provide. Our model is a response to the question, "What can we grow within the confines of our oceans that can sustain both the consumer and the sea?" The answer comes from Mother Nature's own technologies, seaweed and shellfish, designed to mitigate harm naturally.

The infrastructure is simple: seaweed, scallops, and mussels grown on floating ropes, stacked above oyster and clam cages below. The farm was designed with the commons in mind, with an extremely small footprint and light aesthetic impact. Because our farmers don't own the water but rather lease the right to grow in a specific area for five years, the farmed space remains part of the commons and allows anyone to boat, swim, or fish on the farm. The goal of this

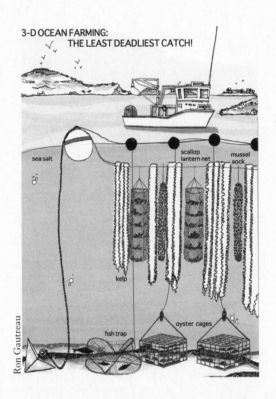

3-D OCEAN FARMING:
THE LEAST DEADLIEST CATCH!

sea salt

scallop
lantern net

mussel
sock

kelp

fish trap

oyster cages

Ron Gautreau

farming model is simplicity, not complexity, and requires low capital costs and minimal skills. By creating low barriers to entry, we hope to foster a new generation of stewards of the commons. No longer pillagers of the ocean hunting the last fish, our model will foster a new generation of climate farmers.

I founded a nonprofit organization, GreenWave, to replicate this 3-D Ocean Farming model and apprentice new restorative ocean farmers in building and growing their own farm businesses. We are creating the market infrastructure needed to ensure farmers capture value in the supply chain, and fostering new kinds of economic relationships among growers and buyers and consumers, through channels like community supported fisheries (CSF). While our CSF provides a secure source of income for farms, access to good local food, and shared community, the vision for the new blue-green food economy has to be expansive. In the era of climate crisis, we have to provide opportunity for community-based food distribution while also considering, on a scale parallel to the enormity of the problem, how to create jobs and opportunities for the sectors that have been left behind. We aim to support the development of hundreds of farms dotting our coastlines, clustered around a seafood hub or distribution center, embedded in offshore wind farms, surrounded by conservation zones, creating a regional "merroir." These networks of farms can have a major global impact on our increasingly insecure energy and food systems, and the vitality and resilience of our commons.

While the commercial fishing industry of the 1980s and 1990s was fueled by a mentality of profiting through pollution with little regard for the future, today, the future is the focus: the future of the ocean, the future of jobs in the blue-green economy, and the future of food. This is our chance to do food right.

FROM HOOKER DAM TO THE NEW MEXICO UNIT:

Making Sense of the Gila River Diversion Project

ALLYSON SIWIK, EXECUTIVE DIRECTOR, GILA CONSERVATION COALITION

The story of the Gila River diversion project begins decades ago, with the dream of generations of southwestern New Mexicans who hoped to tame the Gila's free flow for flood control and use her waters for irrigation, municipal, and industrial use. This dream was an outgrowth of America's policy of Manifest Destiny that encouraged water development as part of settling the West. Federal agencies like the Bureau of Reclamation were established to construct large dams, diversions, and reservoirs that would provide water to fuel the population and agricultural boom of the American frontier.

The idea to harness the Gila started as early as 1911 with the Hooker Dam proposal. Since its formation in the 1930s, the Hooker Dam Association, or the Damsiters, as they called themselves, pushed for "Grant County's favorite dream in forty years," as the Silver City Daily Press reported years later. And in the 1970s and 1980s, they tried three times to dam or divert the Gila but were never successful.

The first real attempt to dam the Gila came about after passage of the Colorado River Basin Project Act of 1968 that authorized the Central Arizona Project (CAP). The legislation provided New Mexico with the right to use 18,000 acre-feet per year of Gila River water in exchange for water delivered to senior water rights owners in Arizona through the CAP. The Damsiters' vision of the Hooker Dam included construction of a main-stem dam across the Gila River just below the Turkey Creek confluence that would have provided water for mining, industrial, and agricultural use, created a lake for recreation, and possibly provided hydroelectric power. However, it would have flooded the Gila Wilderness, America's first Wilderness Area, and it created significant environmental issues. National politics ultimately stopped the project as the Hooker Dam appeared on President Jimmy Carter's infamous hit list of Congressional pork barrel projects, and it was shelved in 1977.

A few years later, as federal laws changed to

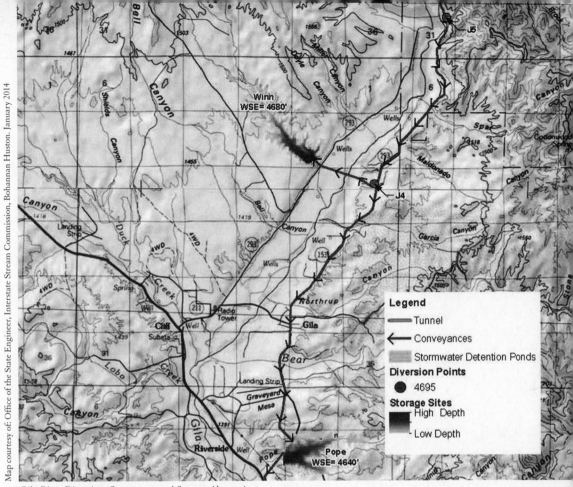

Map courtesy of: Office of the State Engineer, Interstate Stream Commission, Bohannan Huston. January 2014

Gila River Diversion, Conveyance and Storage Alternatives

require assessment of all alternatives to proposed water projects, the Bureau of Reclamation picked up again the evaluation of alternatives to the Hooker Dam through the Upper Gila Water Supply Study. One of those alternatives was the Conner Dam, proposed for a site in the Middle Box on the Gila National Forest. But this second attempt to dam up the Gila also failed because the reservoir would have flooded prime agricultural lands from the Middle Box to the Iron Bridge in the Cliff-Gila Valley and negatively impacted the threatened spikedace.

The third attempt was a diversion and off-stream storage reservoir in Mangas Creek. At the time, this was considered the only Gila River project alternative that could be justified economically. As the story goes, however, after the Bureau of Reclamation spent $6 million in planning and assessment, the mayor of Silver City said, "Thanks, but no thanks," to the project. Indeed, the Silver City Daily Press reported in June 1988 that the Bureau of Reclamation canceled the Upper Gila Water Supply Study citing high cost and no need for project water.

Flash forward to 2015. Southwest New Mexico is discussing the fourth attempt to divert the Gila, and yet the story is still the same: the Gila River diversion project is technically and financially infeasible, there is no demonstrated need for Gila River water, and the diversion will damage New Mexico's last free-flowing river.

The latest effort to divert the Gila River is a result of passage of the Arizona Water Settlements Act of 2004 (AWSA). The Act settled long-standing Native American water rights claims in Arizona. It also created the formal mechanisms by which the Secretary of the Interior may deliver 14,000 acre-feet per year of additional water from the Gila River to New Mexico, originally authorized by the Colorado River Basin Project Act in 1968. The AWSA authorized a diversion of the Gila River, called the New Mexico Unit of the Central Arizona Project (NM Unit), if New Mexico agreed to deliver water to downstream users in Arizona to replace what we take out of the river.

The AWSA provided to New Mexico $66 million in non-reimbursable funding to meet local water needs in southwestern New Mexico without diverting the Gila River. Additional funding—up to $62 million more—was also made available for construction, but only if certain requirements were met. The Bureau of Reclamation has said that it is highly unlikely that New Mexico will receive all of the additional funding. Adjusting for inflation, New Mexico may have to make do with only $90 million from the Feds for project planning and construction.

The AWSA includes the Consumptive Use and Forbearance Agreement (CUFA) that outlines the conditions under which New Mexico can divert Gila River water. Because the CUFA is designed to protect downstream senior water rights holders in Arizona, New Mexico's

ability to divert is highly constrained. Therefore water legally available for diversion is rare. If one applies the CUFA diversion constraints to the period of hydrologic record, water available for diversion occurred less than 10 percent of the time.

Moreover, in order to capture water under these highly constrained conditions, New Mexico is forced to construct high-capacity and hugely expensive diversion, conveyance, and storage

Allyson Siwik

infrastructure that will be infrequently utilized. Because suitable reservoir sites do not exist, expensive lining of off-channel reservoirs to reduce seepage losses is required for the project to function, but the technical feasibility of lining reservoirs is unknown. Even if reservoir seepage losses could be controlled and if climate change did not occur, maximum yield is a small fraction of New Mexico's AWSA appropriation of 14,000 acre-feet per year.

The recently released Bureau of Reclamation Value Study estimates that construction costs for the diversion project range from $800 million to over $1 billion. Although the federal AWSA funding will help offset some of the costs of building the diversion—at best $100 million—it's not enough to cover the shortfall of a $1 billion project and will leave taxpayers and water users picking up the difference, potentially $900 million or more.

Given the low yield and high cost of the diversion, project water could cost thousands of dollars per acre-foot. Still unknown are the intended customers for the water, how much the water will cost water users, and whether the water is affordable.

Unfortunately, since the Hooker and Conner dam days, the Gila's native fish have not fared well. Widespread habitat loss, predation from nonnative fish, and other factors have compelled the U.S. Fish and Wildlife Service to up-list the loach minnow and spikedace from threatened to endangered under the federal Endangered Species Act. A Gila River diversion could make conditions worse for these imperiled fish, as well as the other threatened and endangered species of birds, snakes, and frogs that depend on a healthy river system for survival.

The Gila River Flow Needs Assessment, representing a consensus among fifty scientists from a range of disciplines, concluded that a diversion will negatively impact the hydrology and ecology of the Gila. Analyses of the CUFA show that 79 percent of divertible water occurs historically during the snowmelt runoff season of January through April, a critical time for spawning fish. The physical diversion facilities will segment and damage habitat for listed species, including habitat that has been acquired and protected to mitigate ecological damage for other Colorado River Basin water development.

Since the diversion would be constructed in the wild Upper Gila Box canyon, within the Gila National Forest, an area that has been proposed for Wilderness and Wild and Scenic designation and is a popular recreation area for hunters, fishers, campers, and river runners, the project would have a negative impact on outdoor recreation opportunities.

The bottom line is that a Gila River diversion project is unnecessary. The large Mimbres Basin aquifer underlying the populated areas in southwest New Mexico can meet water supply needs far into the future. In fact, in several locations, local wells have stabilized following past unsustainable use and recovered due to significant agricultural conservation measures. The proposal to pipe Gila River water to Deming would have very little impact on Mimbres Basin water supplies. Although imported water may decrease groundwater pumping, the proposed importation is only about 8 percent of the current agricultural pumping, and 4 percent of the current Luna County groundwater use. The proposed importation to Deming of 2,500 acre-feet per year amounts to approximately 0.008 percent of the volume of good-quality groundwater in storage in the Mimbres Basin.

The common-sense approach under the AWSA is to implement nonstructural alternatives and non-diversion infrastructure improvements to reliably meet community water supply needs at a small fraction of the cost of the Gila River diversion. This is the same conclusion reached by our predecessors twenty years ago. It's more cost-effective, easier, and faster to sustainably manage our groundwater resources through municipal and agricultural conservation, infrastructure improvements, water reuse, and other efficiency improvements.

In fact, the town of Silver City has opted out of the Gila diversion project and has pursued its own course to build a secure and cost-effective water future for the water users it services. Silver City will construct the Grant County Regional Water Supply Project, which will serve 26,000 people in Silver City, local water associations, and the mining district through an intercommunity pipeline and a new well field. It is also implementing water conservation projects and already uses reclaimed wastewater to irrigate the golf course.

Although the New Mexico Interstate Stream Commission, the state agency charged with managing the Gila diversion planning process, did allocate less than 10 percent of available AWSA funding to some of the non-diversion alternatives, moving forward with diversion means that AWSA funding that could be used to fully fund these cost-effective water supply alternatives will instead be frittered away on planning for a fatally flawed project.

At the end of the day, when a diversion project has been defeated, southwestern New Mexico will have no federal funding for critically needed local community water projects, such as the Grant County Regional Water Supply Project or ditch improvements for irrigators.

NEXT STEPS

The local New Mexico CAP Entity is now at the helm of diversion planning efforts, although the Interstate Stream Commission still holds the purse strings, keeping control of AWSA funding and budget approvals. This group comprises thirteen county and municipal government representatives, irrigation ditch associations, and soil and water conservation districts in the four-county area. The New Mexico CAP Entity wants responsibility for the design, construction, operation, and maintenance of the Gila diversion and represents the current board of Damsiters hell-bent on pushing its dream of a Gila River project regardless of cost or technical and environmental feasibility.

Some sense was injected into the process in late November when the Department of the Interior signed off on the New Mexico Unit Agreement, triggering the next phase of environmental and financial feasibility assessment of the diversion project. Interior insisted on supplemental terms being added to the agreement to clarify for the New Mexico CAP Entity what they were getting into by moving forward with a diversion. The supplemental terms will increase accountability and hopefully lead to a robust environmental analysis of the proposed diversion. The important gains include financial feasibility and accountability, that projects must be designed according to federal standards, that less money may be available for diversion, and that the full range of environmental compliance is necessary. Finally, the Secretary of the Interior can make an independent decision that is best for the environment.

If the environmental compliance process is honest and rigorous, it's highly unlikely that a diversion would be built, due to huge costs, technical infeasibility, and damage to the river and the seven endangered species that depend on it.

Unbelievably, after a decade of study and $5 million spent, the New Mexico Interstate Stream Commission and the local New Mexico CAP Entity have chosen to move forward with a diversion project even though all indications tell us that this project is doomed to fail. It's a shame that those responsible for the diversion planning haven't learned from this project's long unsuccessful history. Our best hope is that common sense prevails soon and what's left of

the AWSA funding can secure a reliable water supply for everyone in southwest New Mexico.

Editor's note: For most of you reading this, the Gila is not your river, but do you know the story of your river, of your watershed? Its industrial history, the major users, the pollutants, the insects, the aquatic mammals? This information is relevant to the future habitability of your chosen place, and in-depth knowledge, well shared, can change the culture of belonging in ways we may not fully appreciate, or at least not at first. Listen to Alysson's talk at the OUR LAND 2: Tracing the Acequia Commons Symposium, New Mexico, November 2016: www .agrariantrust.org. Find out about her organization, GRIP, and listen to her radio program, *Earth Matters*, which broadcasts on Tuesdays and Thursdays at 10 a.m. and 8 p.m. via webstream or anytime via podcast at gmc.org. —SvTF

Allyson Siwik

❀ ❦ ❧ ❁ ❊ ✳ ❀ ❋

In proportion as our inward life fails,
we go more constantly and desperately
to the post office. You may depend
on it, that the poor fellow who walks
away with the greatest number of
letters ... has not heard from himself
this long while.

—Henry David Thoreau

FEBRUARY

DRY LAND

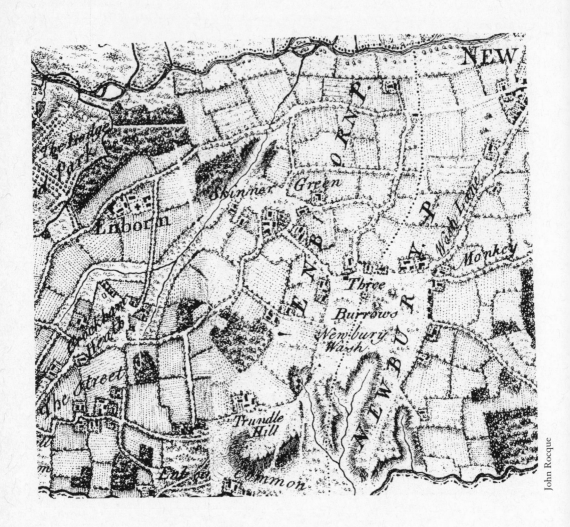

John Rocque

THE ELEPHANTIDAE, THE FLAME, AND THE CHAINSAW:

Wildfarm Release

NIK BERTULIS

Editor's note: The opinions in this piece are real and unmodified. If you disagree, Nik would probably love to debate with you, or you can duke it out with physical evidence of land management. So have at it, friends, and may the debate bring lively ecosystem benefits to those around you. —SvTS

Of all the animal kingdom, we mammals need grasses the most. From our cereal grasses, corn, wheat, oats, and barley to the meadows and pastures that make the best butter and cheese. We swung from our arms out of trees onto grassy savannas on our feet, the birth of bipedalism. Science tells us humans and mammoths shared the North American continent for about six thousand years. This was the end of the Pleistocene, a period that lasted from two million years ago until about eleven thousand years ago, well before the advent of agriculture.

Indubitably, this was an exciting time to be alive and a time we shared with ten times the number of megafauna we have today. Salmon nine feet long, fourteen species of pronghorn deer, six-hundred-pound saber-toothed cats, and condors with a sixteen-foot wingspan, all cohabiting together in our own Serengeti. This epoch was truly a heyday for mammals, not just in North America but globally. The sheer magnitude of animal biomass dwarfed today's vast and biologically vacuous agricultural monocultures.

If you have a Web device handy, search YouTube for "elephant knocking over tree." Now imagine that scene repeated billions of times over millennia, and you will see how the Elephantidae earned the title "kings of the savanna." Paleo-ecologists describe how mammoths and mastodons, like their contemporary descendant, the elephant, push trees over and gorge on the tender new shoots and grasses. The result is a powerful cycle that initiates a unique ecosystem. African ecologist Norman Owen-Smith calls them the shapers of landscapes of plenty. Extending this pattern to American landscapes, these

85

proboscideans would have created photosynthetic boomtowns for the acorn-chestnut-berry-tuber and grass polycultures where wild mammals (and their domestic proxies, pigs, goats, sheep, cows) and people could thrive.

With hooves, dung beetles, and bazillions of soil microbes, we mammals "plowed" trillions of tons of dung and urine into the soil. As any organic farmer knows, this is a potent recipe for a fertility that yields abundant harvests. The threat of predation by the smilodon, dire wolves, and short-faced bears kept ungulates—and probably us *Homo sapiens*—bunched together for protection, which increased the tillage-like effect of hooves on soil—hat tip to Allan Savory.

Postulations of the whys and hows of the extinctions of North American Elephantidae are many and fiercely debated. But once they were gone, the human response to their absence was obvious: only fire could replace the ecosystem function of our elephant king. Ask Karuk Nation members Bill Tripp or Frank Lake, whose people depended on the Karuk oak savannas for millennia, about fir encroachment. The relationship of these fire-adapted and dependent ecosystems to the pre-European peoples that lived across the Americas is fascinating.

Today, thousands of land managers and game wardens across the world practice prescribed burning for similar reasons. The grasses come back strongly, with adapted native species, reduced wood load, and healthy forage for wildlife. In the Southwest, ranchers even bulldoze and chain the pinon and juniper to mimic the action of fire on the landscape.

What might today's farmers, gardeners, and land managers glean from all this Pleistocene ecology, especially those keen on decoupling from the fossil carbon grid? Well, we can selectively harvest our woodlots for biomass to run wood gasifiers for heat, power, and char production. Grazing animals on grasslands could permeate these woodlands. Herds of grazers, both wild and domestic, pooping atmospherically derived carbon into and onto the soil is a potent tool in fertility management and carbon drawdown. Soil scientists Ratan Lal and Piet Buringh talk about having lost somewhere between eighty and five hundred billion tons of soil carbon to the atmosphere in oxidation. Imagine the return of the great herds to the earth's billions of acres of degraded grassland ecosystems: according to Allan Savory, by some calculations we could bring atmospheric carbon down to preindustrial levels in forty years.

The chainsaw, growling icon of environmental destruction, becomes a proxy for the elephant's tusk—helping us strike a better balance between the forest and the grassland. Ecological restoration thus becomes the key link to both food security and the maintenance of biodiversity.

In forestry there is a useful term for this process: release. The preferred trees are released from competition for nutrients, sunlight, and water. So called weedy trees are removed for lumber purposes—but there could be other motives in the forest. Douglas fir trees, which reach four hundred feet tall in parts of the West, can and will shade to death the oak savannas that provide a cornucopia of food for a tremendous array of species. The tallest oaks, by comparison, grow up to one hundred and fifty feet tall. Imagine if we were managing for the oaks the kind of abundance that would be possible in a forest full of nuts. I wish not only for oak release but also chestnut release, hickory release, hazel release, and pecan release.

We *Homo sapiens*, grass junkies that we are, cleared all but less than 1 percent of Eastern primeval forests. In the West we are doing a bit better, with 4 to 7 percent, respectively, for extant old-growth redwood and Douglas fir forests, and yet they are still diminishing, as the architectural appetite for these ancient woods is fierce.

We need forests, both young and ancient, for beauty and timber, as do countless species of birds, fish, insects, amphibians, and other mammals. If we humans and our nonhuman friends are to survive and thrive, we need stewardship that is grounded in stories of the First Nations as well as stories of science.

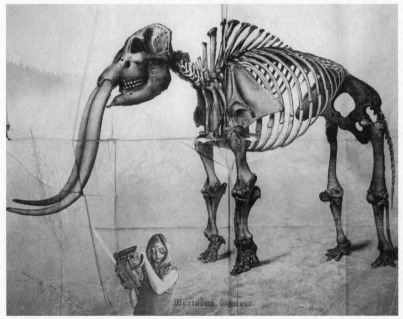

Mastodon Giganteus.

Nik Bertulis

We need stewards of the mosaic of forests and grasslands, polycultures wild and domestic, for, like it or not, by the undeniable impacts of our tremendous profusion, we are the new kings; we might as well be good kings.

Imagine a day when our children can walk out their back doors and, depending on their mood, choose to venture into a field of mixed vegetables, a food forest dripping with falling fruit, a grassland where the ancient drama of predator and prey unfolds, or an old-growth forest where they might practice what the Japanese call *shinrin-yoku,* "forest bathing."

Learn to use a chainsaw, grow old-growth forests, learn your grasses, study predator-prey relationships, see every poop as another half pound of carbon moving from the sky to the soil, and for the love of Pachamama, hug a regenerative logger.

FARMING PEACE

JOHN JEAVONS AND MATT ANDERSON

The fate of an individual or a nation will always be determined by the degree of his or its harmony with the forces and laws of nature and the universe.... The fullness of life depends upon man's harmony with the totality of the natural cosmic laws. Our individual evolution is a job that has to be carried on day by day by each individual. It is a lifelong task.
—Richard St. Barbe Baker

My mentor was an accomplished thespian and master gardener named Alan Chadwick. An apprentice of Rudolf Steiner, he developed his own gardening method that he called the Biodynamic French Intensive Method. He was a wonderful and complex man with one resolute goal: to end war.

Like any veteran of World War II, Alan experienced a tremendous amount of death and suffering. He went to South Africa shortly after the war, where he met Freya Von Moltke. Freya had fled from Nazi Germany after her husband was killed for his involvement with the peace movement. Together Alan and Freya decided that they would work to end war, and their chosen method was gardening. They believed a culture concentrated on nourishing their local habitat would prevent the conditions that drove people towards large-scale industrial warfare. And, as people began to breathe life into the soil, they would also breathe life back into themselves.

Prior to meeting Alan I had been on a personal mission: I wanted to discover the smallest area of land in which I could grow all of my own food, clothes, and income. The plan was to develop an environmentally sound and equitable farming method that could meet the basic needs of everyone on earth in the most efficient manner. I felt that if the world used this approach, or a similarly effective one, everyone could live well.

I packed up and drove to the Central Valley of California to interview those who could answer my question: farmers. When I asked how much land I needed to grow my whole diet, nobody had an answer. Many of them scratched their head or glanced upward as though they would receive some insight. One farmer did give me an answer: he told me if you grew a thousand acres of wheat, you could produce enough income to break even—if it was a good year.

This experience helped me to realize that the way we farm does not have a fully holistic foundation in the way it feeds people. I began to see that feeding people had become a byproduct of an industry mostly concerned with the efficiency of profit maximization. Large-scale farming, monocropping, machines, and toxic substances have gotten us away from the reasons we are growing food in the first place. Modern farming has reduced our agricultural workforce to less than 1 percent of Americans, and the demands of the market have kept the wages and prerogatives of many of those people in a state of poverty.

But this is nothing new; the issues of income, surplus, and poverty have been with us since the dawn of agriculture. In fact, they have always been an inherent feature of civilization. Our current situation is just an evolution of the decisions made by our ancestors and reinforced until today.

I did not recognize it at the time, but hidden in the question I was posing to these farmers was a challenge not only to the food system but to the foundations of our way of life.

OUR SOIL IS DYING

For at least ten thousand years we have had a one-way relationship with the soil beneath our feet. With increasing assertiveness we have mined the nutrients from our soil, destroying the fragile ecosystem that Nature so patiently built.

Until now, this has served us well. We have wrenched such a surplus from the ground that we could multiply and fund our civilization's many achievements. Because of soil we have been able to build pyramids, ships, and commerce. Soil is directly responsible for the paper you are holding at this very moment. In fact, you could argue that the real "fuel" that propelled us to the moon was ultimately soil itself.

We have been able to do all of these things because of the rich reserves of organic matter and minerals in our soil. But these reserves are rapidly disappearing. Desertification now threatens us all, and by some estimates there is as little as thirty years of farmable soil left on earth. Our relationship to the soil has allowed us to thrive in extraordinary ways. But we have failed to recognize the responsibility inherent in this relationship. The soil is now overworked, depleted, and susceptible to illness. As a result, we are becoming increasingly malnourished and more susceptible to illness.

A clear way to visualize what we have done is to imagine our soil as though it were a cow. For ten thousand years we have been enjoying nourishment from this cow, and yet we have not been properly feeding it. As a result, the milk that our cow produces has become thinner and less nourishing, weak, and prone to harbor sickness. As our depleted cow has been able to produce less, we have become hungrier. To energize our cow to produce more milk, we have stimulated it with additives in the form of petrochemical fertilizers. And to reduce the sickness that plagues our milk, we have fed the cow poisons.

Our once healthy cow is now sick, and it is dying. It is not providing us with the nourishment we need, and yet we do just about anything except fully replenish its food.

It is not just an abstract or poetic notion that we are linked to our environment. We have been stripping the health of our soil to the point where it can no longer nourish us. We have to fundamentally change our relationship with the soil. Only by replacing the organic matter we have taken can we hope to have truly healthy human beings.

In Gandhi's words, "To forget to dig the soil is to forget oneself."

TOWARD A LIVING SYSTEM

Humans have shown folly in treating only the symptoms of our agricultural problem. Disease and pests are largely absent in healthy ecosystems. Soil provides an abundance of food when it is replenished with the nutrients it expends to grow that food. When we grow a significant amount of

Sarah Gittins

our crops for compost to feed the soil, the results are spectacular. Soil fertility leaps back, pests are almost nonexistent, and the yields can be beyond compare.

I am constantly amazed at the power of Nature's living systems. Over time, I have become sensitive to the inherent restorative capacity of life, and we are working to understand it better. Through observation and careful adaptation, we are developing a method based on successful millennia-old approaches. This method helps the soil and the food it produces thrive, so we may feed ourselves sustainably. In doing this we are directly witnessing that working with life creates more life.

Nature is a wonderfully complex web of relationships. It dances and sings. There is harmony. We have only to join in this process by sharing our care and our labor, and we will rediscover a powerful and generous ally.

TRANSFORMING A FEELING OF ANGER

Many of us work tirelessly, converting our time and labor into a comparatively artificial substance to gain our basic needs like food and shelter. We learn to create a surplus of this substance so that when we age, the system will continue to provide for us on the reserve we have secured. With potentially as little as thirty years of farmable soil remaining, the longevity of this way of life appears unsustainable.

For this and many reasons, we see a restlessness from urban citizens worldwide. People everywhere are becoming aware that something is fundamentally wrong, and a growing amount of anger is being unleashed in protests, demonstrations, and the attempts to stop them. While this is an exciting and necessary step toward changing our current trajectory, we must not let anger carry on in this form. It will lead to war, and it will only intensify our problems, pushing our society and our planet closer to a situation incapable of supporting life.

Instead, we can transform these feelings of anger into joyful collaboration that benefits the earth and our soil as much as it benefits us. We have the exciting opportunity to begin investing in natural systems that will allow us to transform increasing scarcity into abundance. Working with the living systems of Nature, we will find exciting and rewarding employment that provides our essential needs and a meaningful way of life. In the process, we can become participants in bio-remediating our damaged ecosystems.

The path to peace in the world is a path first of peace within ourselves. War will only end when we first end it in ourselves.

Inside every one of us is a garden, and each practitioner has to go back to it and take care of it.… You should know exactly what is going on in your own garden, and try to put everything in order. Restore the beauty; restore the harmony in your garden. Many people will enjoy your garden, if it is well tended.
—Thich Nhat Hanh

FARMING LITERACY

Our other primary challenge in creating a sustainable regional food system is farming literacy. Since I began this work, the majority of the world population has become urban. As people have migrated away from tending the land, we have lost an alarming amount of farming knowledge. Just think of the rich diversity of language, culture, and skills that are evaporating in our time. At this moment in history we need to decide if we are going to regenerate our farming literacy or risk losing it. Only people can maintain this living knowledge, and once this chain is broken, it is gone for good. When a variety of plant fails to be sown, a picture of it will not feed us.

So this parallel combination of depleted soil health and decreased farming literacy presents us with a simple but exciting challenge: we need to cultivate a new paradigm. As we regain our farming literacy, we will build new skills and form new relationships with each other and with Nature. The communities that share information and educate others in living sustainably will become the new cultural oases. As we have more working demonstrations of how a new paradigm can prosper—despite growing scarcity—we can replicate what works and rapidly switch directions for humanity. Never before have the stakes been so high, but never before have we have the knowledge, tools, or potential to act so quickly.

Each of us has innate access to the insight and knowledge that allows us to thrive. Through design, we can ensure that it is sustainable. And together—through responsibility for the needs of our community, those of the earth, and our own—we can create a world in which it is great fun to live.

Farming is not a job. It's not a vocation. Farming is a way of life. When we reorient our relationship to the land so that the goal is sustenance and life, we will be on a path toward a sustainable, abundant future.

GROWING PEACE

When I met Alan in his beautiful garden at the University of California, Santa Cruz in 1970, I got the sense that I had found the trailhead for my life's work. His method was so beautiful, and the results were more than apparent; he took depleted, overgrown plots and turned them into a lush oasis. I saw that it was really possible to remediate land through this biologically intensive farming method, and I knew at once that Alan's method was the key to answering my question.

For the forty-four years since then, I have devoted my life to the development of the Grow Biointensive method. My friends, colleagues, and I have been determined to identify and expand on the scientific principles that undergird this biologically intensive food-raising. Our research has drawn from a rich reservoir of knowledge developed over millennia by a wide variety of cultures in the Middle East, Africa, Asia, and Latin America. This understanding can play a major role in the development of a fully sustainable food system.

To bring this system to fruition, people everywhere need to become involved. We can responsibility of feeding ourselves and our communities. We must each play a role in transforming the non-optimal features of the current paradigm as we repair our depleted environment. I believe we have the opportunity to fulfill Alan and Freya's vision of growing peace. Working in harmony with the life force of the planet, we can cultivate a new human culture based on care for the earth.

The whole world is a garden. And what a wonderful place it would be if each of us just took care of our part of the garden.
—Voltaire

R. de Salis

Editor's note: Here are some groups working in Palestine around food and farming: Bustan Qaraaqa, Tent of Nations, Palestine Heirloom Seed Library/A.M. Qattan Foundation, and Hida Refugee Camp Rooftop Gardens. You can go there as a witness for peace and to learn more about the peace movement, peace practices for activists, and lessons for conflict management. —SvTF

TO FARMERS AT DINNER

A meditation on farm-to-table

ALI KRUGER

Will you sit at my table, and break bread with me?
I long for your stories to sweeten my soup,
A political rant to spice my meat,
Your full-body laugh to melt my heart.

Will you linger in winter's pause to gather around?
I'll listen to your dreams of gardens and greens,
We'll sing to the freezer hibernating your seeds,
You'll smile at the way things always work out, different than we think.

If the moment is evening, will you plant here at my hearth?
From your palm into the ground, from soil to soup pot, from earthen bowl to your lips,
We feast on popped corn, spinach pie, squash stew, baked apples,
The harvest we save becomes the harvest we savor!

Thank you to,
Thank you to,
Thank you to the farmer, for this food we eat!

Thanks to you,
Thanks to you,
Thanks to you, dear farmer, dreamer, doer,
At last we meet!

THE DREAMGOATS

LEAH SIENKOWSKI

As a field hand in the Midwest, I have no land of my own, and my work disappears under snow for several months of the year. Living in an urban center and commuting to the outskirts has been an education in the relationship between conservation and culture, history and technology, domesticity and wilderness. I relish the dissonance, yet I'm torn—between advocating for farmers and working as one.

Last winter I drove west, as young landless farmers must, in search of sun and some unfrozen soil to dig my hands into. Incidentally, I found myself in the high desert of New Mexico, working off the grid and without running water, alongside a herd of free-range dairy goats. The rich history of goats

WEISSHORN
Oh let thy heart make melody,
And thankful songs uplift,
For Christ Himself is come to be
Thy glorious Christmas gift.

quickly intrigued me—in their ten-thousand-year existence as domestic animals, they've provided sustenance to nomads, sea voyagers, and desert smugglers. Not only do they persist in harsh environments, they transform them into milk. As I followed these goats down rocky highways, over cliffs, and through arroyos, I began to learn the varieties of yuccas, cacti, and piñons, which were at first alien to me.

Historically, the life of a goat-walker was isolating, difficult, and irregular—but also steeped in simple, tranquil beauty. Nomadic pastoralism was certainly a matter of survival, not theater—animals and caretakers tucked away in the farthest countryside, always moving on to the next piece of land. At the dairy in New Mexico, the rhythm of our lives lined up with the needs of the goats. Our days were bookended by milking, punctuated by feeding and, less often, birthing and slaughter. We ate thick, creamy chèvre made from their milk; we relished their hearts and their livers with onions; we peeled the meat from their ribs.

JUNGFRAU

Great is the mystery
Of wondrous grace,
God manifest we see
In Jesus' face

O deepest mystery
Of love Divine,
God manifest for me
And Jesus mine!

I resonate with the scrappy ingenuity of goats—their spunk that carries into the taste of their milk, their disbelief in fences. My dream is to take them on tour, giving goats the stage they are practically built for, while showcasing a better, ecological alternative to lawn-care chemicals and powered mowers.

With goats, I can bring agriculture into places that seem inedible or useless. I can bring attention to unused lots and bits of overgrown land as much as I can for farmers themselves— all while wearing down hoofs against cliffs of pavement, reshaping expectations, and advocating for surprise. Without land of my own, I can be a modern goat-walker, my hands engaged in both advocacy and farming, working to transcend the urban-rural binary and rewrite our ecologies to include grazers and browsers once again. Thinning the overgrown and adding fertility to the forgotten, I can reimagine a world that is hospitable to farmers, connected to its food, and interested in ecology. Turning brambles into milk, I can contribute to making the world a safer and more edible place.

USING SUPPLEMENTARY CROPS FOR FEED AND RENOVATION

JOHN SNIDER

Organic farms in the West are learning their most available source of fertility is sunlight; it's free! A mind-boggling array of crop fertilizers in the organic sector has not instilled confidence in their use, let alone their efficacy. More needs to be done, certainly, to monetize these products. Meanwhile, photosynthetic conversion of green growing plants and living roots into money via the rumen of livestock is taking hold once again.

"Green manure" and "soiling," two terms used to define a then-mysterious process, were an act of survival in eighteenth- and nineteenth-century England. Turnips, a member of the Brassica genus, were grown as fodder and cover crop. Food production increased 300 percent. The weak and sickly ley system—one year of wheat to pay the rent, then three or four years of grass to recover—was reinvigorated.

The turnip, they say, instigated the industrial revolution. More food and higher birth rates meant emigration as well. The turnip and rutabaga followed the diaspora to Australasia, where warmer winters and drier summers meant new innovations in plant breeding, specifically forage brassicas. Enter the taproot. Forage brassicas not only survived hotter, drier weather but offered extended grazing over multiple seasons.

Green Springs, in California's San Joaquin Delta, has a grazing crop system for its integrated organic farm. Last year, after a tomato crop, Winfred Brassica, Graza Brassica, and Tonic Plantain, in a "relay" crop, was planted to cycle nutrients and use more sunlight. The days of idling $15,000-per-acre California farm ground for the winter are over. After all, this is the land of sunshine.

Editor's note: John Snider specializes in rare varieties of field plantain, turnips, parsnips, chicories, and other deep-rooted plants, which are grown more commonly in South America to remediate compacted soils. He has an intensive approach. Up in Colorado, in the San Luis Valley, a young and older rancher team is working with ranchers to graze down crop residues, feed out corn on potato fields, and even plant in rotations to be grazed as a way to increase the territory for grazing and to fatten grassfed animals

We estimated the dry matter production to be ten thousand pounds per acre at first grazing. Frank Savage, a lifelong cow and sheep man, knows his production well, as he is paid on the gain. He calculated 8,700 pounds was consumed by a crop of lambs. The rest, in manure, urine, and all the green carbon exuded through the living roots, is converted to food for the soil.

He has good cover and plenty of residual to promote multiple grazing. This capitalization, if you will, of regrowth converts to profit per acre; just ask him. He knows animal performance has a direct relationship with the crop residue. He is also converting the fertilizers left over from the tomato crop into a food source for all the creatures, great and small.

His next crop is already in the ground. The relay concept saves time, money, and undue tillage. His Winfred and Graza are nurturing ryegrass, Gala Brome, clover, and Tonic Plantain. He expects this crop to supply him with more feed of higher quality for longer, into next year.

Forage brassicas, progeny of the common turnip, and their magical world-changing properties are taking the confusion out of soil correction and crop fertilization. Grow your own.

John Snider is PGG Seeds' Oregon-based agronomist. He provides expertise on forage and pasture systems, ensuring the best match of seed cultivar to farm system, soil type, and climatic conditions to maximize productivity. For more information on PGG seeds, contact John at john@pggseeds.us.

throughout the year without importing hay; this also improves the health, biology, and tilth of the plowed and irrigated croplands, which benefit tremendously from this "animal action." Martha Skelly and John Kretsinger presented on this last year at the SWIGLA conference. —SvTF

Clem Powell reading on a boulder in the Colorado River

THE PERILS AND PROMISE OF WESTERN WATER POLICY

HOWARD WATTS III, GREAT BASIN WATER NETWORK

Arid lands are special in the lexicon of spaces, defined not by abundance but by scarcity. Despite its rarity, water is the pulse of these places. As rain, water carves canyons with flash floods that put weathering into overdrive. As snow, it feeds streams and lakes year-round through gradual melting, sustaining a surprising richness of life as a result. Water seeps underground and appears miles and years later at springheads, creating wetlands and oases. With a little water, the desert explodes in the myriad sounds, sights, and sensations of life.

The Great Basin is a particularly fascinating arid region, centered on Nevada but also extending into Oregon, Idaho, Utah, and California. Long sagebrush valleys surround islands of mountain forests in neat north-south rows. Rivers drain not into the ocean but instead to lakes and playas within the regional sink, hence the Great Basin name. Home in the not too distant past to glaciers and massive lakes, it's now the driest part of the continental United States. Most of the land is owned by the public, and its diverse resources are subject to often competing uses, such as agriculture, mining, urban development, rural communities, indigenous peoples, and stewards of nature. Central to all of these human conflicts is water, but because the written law has superseded natural law and water is treated as a commodity instead of a commons, every creature in the Great Basin now faces an existential threat.

How did we get to this point? Just like the degradation of any public good, there are many decisions that made sense individually but have collectively built into a major problem. The creation of the beneficial-use doctrine essentially flipped the natural notion of water use on its head. All nonhuman water use became waste, and the goal became to use every last drop to advance the goals of human settlement and enterprise. Combined with a growing ability to engineer water away from its origin and a failure to understand how interconnected, fluctuating, and ultimately fragile the hydrologic system is, we've accumulated water debts that we will never be able to repay.

The Great Basin is the birthplace of many of the great engineering marvels and human follies of the arid lands. The Newlands Act created the U.S. Bureau of Reclamation and led us to believe that with enough determination and money, we could provide water anywhere. About a decade later, Los Angeles would snatch up water rights and build an aqueduct to drain Owens Valley; city ratepayers still foot the multimillion-dollar bill for the mitigation of the toxic dust bowl that resulted. About fifteen years after that, Hoover Dam was finished, using hydropower to pay for reclamation and reservoirs to store water "lost" downstream. As dams and diversions multiplied, we also learned how canyons drown, lakes evaporate, rivers die, and the thirst for more only grows.

Now Las Vegas wants to build a three-hundred-mile pipeline up the Eastern edge of Nevada, ending at the valleys hugging Great Basin National Park. At a total cost of over fifteen billion dollars, it would take the "unused" water of rural towns, ranchers, and Native American communities and pipe it down to the city, dropping water tables, killing plants, and turning increasingly rare pristine Western skies into a dustbowl. While we fight in court and educate the public, this public agency runs a subsidized ranching operation to advance its control of the area's land and water resources. Disastrous culturally, economically, and environmentally, this growth-greedy hyper-commodification and long-distance manipulation of water is a way of thinking that should and must be left behind.

The solutions are simple in theory and difficult in practice. We must first commit to the goal of getting every region, sector, and creature through the crisis of water depletion in the arid West. This breeds the collaboration, creativity, and goodwill that are required to meet the challenge. Next, we must set a carrying capacity for the land, determining how many of its various inhabitants we can support on the current local water supply. While undoing some of our disastrous plumbing seems unlikely, we can start by ensuring that no more is built. We must amend our laws and budgets to make conservation a top priority and treat the water system holistically. Like climate change and the other environmental challenges we face, there's no going back. But, by acting boldly, quickly, and together, we may be able to preserve the water-dependent land as experienced by those living today for the benefit of future generations.

MARCH
WET LAND

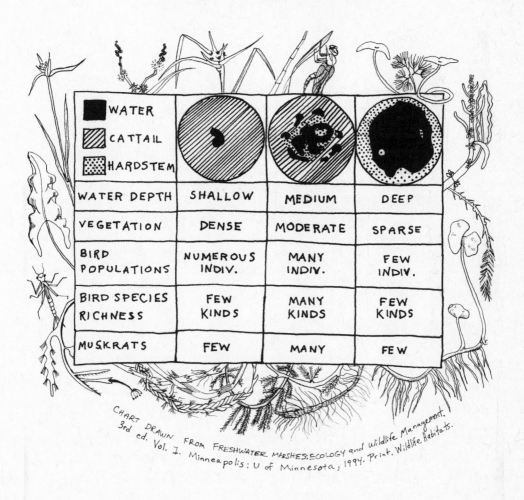

	WATER	CATTAIL	HARDSTEM
WATER DEPTH	SHALLOW	MEDIUM	DEEP
VEGETATION	DENSE	MODERATE	SPARSE
BIRD POPULATIONS	NUMEROUS INDIV.	MANY INDIV.	FEW INDIV.
BIRD SPECIES RICHNESS	FEW KINDS	MANY KINDS	FEW KINDS
MUSKRATS	FEW	MANY	FEW

CHART DRAWN FROM FRESHWATER MARSHES: ECOLOGY and Wildlife Management. 3rd ed. Vol. I. Minneapolis: U of Minnesota, 1994. Print. Wildlife Habitats.

GOAT ONION SOUP:

A Recipe

JASON BENTON

SPRING

It begins in early spring when the green onion sprouts emerge from their winter slumber up through the rotten compost and manure. How do they do it? The bulbous plants—onions, tulips, garlic, daffodils, crocus. Every spring I watch their green shoots rise from their hidden bulbs straight up through the tree leaves and plant detritus that has been pancaked to the ground by a few months' snow cover. The holes they cut through the leaves seem to magically dematerialize, making way for the shoots. There is no sign of struggle or that the emerging bulb shoots displaced the leaves in any way, but it is as though they surgically separate the leaf matter at an atomic level a millimeter ahead of their growth.

This is one of many wonders I choose to accept as is and just wait every spring to behold it. I won't let Google ruin it for me. Awe and effective contemplation erase ignorance and beget deeper thought and—most importantly—feeling, allowing for broader and more profound possibilities of experiential understanding. Sure, they're fun and interesting, but scientific facts as such are sterile, cheap, and easy, and

they simplify the surreal. They abruptly halt feeling and thought, turning what could be an act of communion and relationship, which are complex reciprocal, into a passive, idle non-act of consumption: "[Burp!] Next."

The melting snow makes everything wet, cool, and musky, and this too lends an emotion. If you've ever had cabin fever, you know what I'm talking about—the sensation of the winter chill in your bones finally thawed by a higher and warmer sun, even if the sun remains hidden behind clouds. Your mind is oblivious, but your body just knows it's time. The birds feel it too and sing their spring songs and begin building their nests. The children run outside without jackets (my daughter throws her boots and socks off), even though it is just over forty degrees Fahrenheit, the ground is yet frozen, and there is still enough ice to skate on. Do the onions feel this too?

If so, the sage doesn't. It sprouts several weeks later—it needs more sun and more heat. As spring turns to summer, sage's leaves seem to collect the sparkle of the sun and begin to grow silvery. When all else wilts under the long dry summer days, sage sings with joy. Sage needs space and sun, not much else. Onions need more—space, compost, water, and attention. Throughout the summer I pick a sage leaf to chew on my way out of the garden—a warm sparkle of sun in my mouth.

Or perhaps the soup began last year when I planted the hops for the beer? Or a few years ago when I started the heirloom onion patch and compost bed? Or when the maple sap began to flow in late February—the sap which I used to brew the spring beer—and fill the sap pots? At about the same time the onions were emerging, the primrose was reaching bloom, and the pine trees had new soft green nubs on the ends of their branches. The fresh primrose and pine nubs were used along with the hops (dried and packed the year before) to bitter the beer, lending notes of mint, herb, citrus, and pine.

When does a good soup, a good life, begin and end? Within minutes after this soup passes through me it will be absorbed by the oak trees whose roots have penetrated my sewage leach bed. And quickly after that, the water will be reconstituted into the atmosphere by both the tree and me.

SUMMER

A goat wether born around the same time as the onions sprouted will grow by, and with, leaps and bounds over the summer. The goats seem to come alive with the spring just as all else. In April they will be bouncing and basking in the sun. By May they are like shop-vacs, inhaling all the green in their path. The wether eats voraciously all summer long, keeping his rumen more often than not full. Sassafras leaves are goats' favorite. In the cool summer evenings I'll bend a young sassafras tree over, pulling down the high branches for an evening dessert. The boughs will spring back up and sprout out new leaves after a few weeks. The goats watch my movements and eyes carefully—they know when I am looking for a good bough to pull down for them. The wether will swallow the leaves whole as fast as he can, filling his rumen taught like an overfilled balloon. He knows that pausing to savor the first bite will cost him twice the pleasure later. In the evening, the area where he beds down for the night is incensed with the aroma of sassafras tea as he burps his cud with smiling eyes. I sit with him and watch the red screech owls pester roosting robins. When August approaches, I make sure he gets the lion's share of sassafras leaves, and we continue this routine all the way through October, when the leaves turn brilliant shades of red, pink, yellow, and orange.

You ever sit with a ruminating ungulate? It puts you into a dreamlike, slow contemplative state. Sitting with my goats, smelling their cuds, thinking about nothing, I realized I am next to nothing without a rumen. And my gas is not nearly as sweet as a goat's. You can't eat well without fermentation. My rumen: the carboy and crock, the fermenting yeasty rising bread dough, the meat I eat is first aged—fermented, rotted. The most flavorful foods have some sort of microbacterial process—cheeses, krauts, breads, alcoholic stuff, pickled stuff, buttermilk, sourdough, sour cream, sour everything, and yes, beer. Now I know why ungulates seem to browse like it is a chore, but take their time to relish the cud. This is why the goats compete to fill their rumen with the sweet sassafras, then find a comfortable place to burp the cud. I've even watched one goat sniff another's cud burp as if he was checking out what is for dinner. A goat will meander about and do other things, perhaps look for better food, if her cud is boring; but if it is delicious and worthwhile, she will lie down in a comfortable place to savor it.

A late August thunderstorm blows down a centuries-old red oak. The girls and I, drenched and squealing, just shivered in from dancing in the warm rain when it landed on the power line. As the power line grounded on the neighbor's barbed-wire cow fence, it sent a miniature lightning bolt out of each barb along the wire. The township workers came quickly with a backhoe to clear the road for the utility crew. I asked them just to push the tree into my yard; it'll make great firewood.

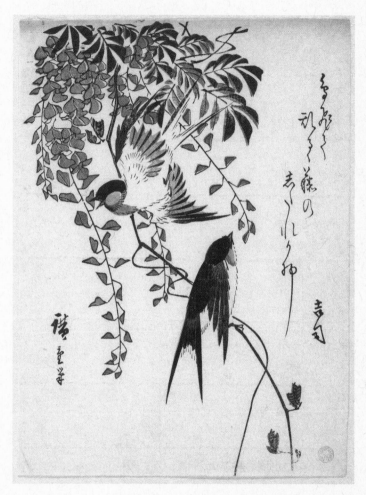

The tree was so huge it needed to be cut into several pieces before it could be bulldozed off the road. Family and friends came to help clean up the branches while I recovered from an injury. It took me the rest of the summer to split and stack, maintaining a constantly calloused hand.

Once the township crew left, my neighbor the old cow man came by to fix his fence, and I helped him cut a few posts from the fallen oak branches. As we mended the fence, he complained that all his kids and grandkids were gone, moved to town or in college, no one left to help bring in the hay. A horrible hay year anyway. "But they have good jobs comin', hain't that somethin'?… Hey, can ya help bring in the hay?"

I split the wood and stacked it along the sunny side of the barn lean-to. It dried quickly there, filling the barn with the spicy scent of oak, reminding me of a ski lodge I had visited last winter and foretelling the cold to come. Nothing smells like oak. You hold the scent in your nose and taste a warmness not unlike hot pepper, fresh bread, and ripe fruit. And smells resurrect memories and their associated feelings so well. Oak: boyhood memories of a sawmill, childhood memories of a babysitter's toasty fireplace, the heavy oak beam I raised in my house with the help of family and friends five years ago, many New England ski lodges, stacking wood in the fall and all the autumnal colors, fabrics, and sensations that come with it. The girls played, stacking the wood like blocks into castles, bridges, and Rapunzel towers. Within a day a nest of gardener snakes and a baby groundhog found refuge under the pile and the goats made home on top. The goats quickly diminished our neat cord stacks into a messy heap. Should've known.

FALL

The onions are allowed to bolt, and once all the seeds have dropped, the onions wilt, and they are picked and hung in the cellar to dry. They are extra pungent this year. As I leave the garden with a bucketful of onions, my wife asks me, "What's wrong?" The onion seeds are spread with a rake and covered with composted barn straw and manure from the chickens and goats. Some of these seeds will make autumn green onions and the others are left alone to lie dormant until they sprout again next spring just as the last snow melts away.

After the first hard frost, the sage is snipped a few inches from the base and hung inside to dry. I put the sage stumps to sleep under straw for the winter. They will sprout back up next spring. The back yard is now full of the spiritual aroma of sage, and I can't help but put a large leaf in my mouth to suck on. Fat fall rabbits are trapped in the garden fence, and I think about poaching one for a stew. All summer long the rabbits are as tame as a lap dog, lazily chewing clover and dandelion, but at the first wisp of cool autumn air, they are full of panic and dread. How do they know? The shorter days, crisp cool breeze, and rustle of fall leaves lend to me a nostalgic feeling of festivity, warmth, and an appetite for

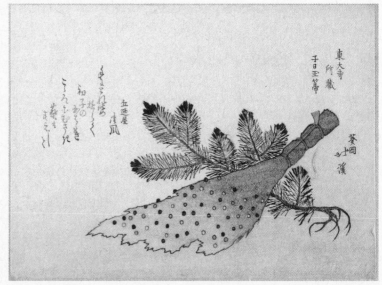

pumpkin, warm drinks, and … rabbit. And to the rabbit, fear. Once the sage is dry, it's kept in a sealed container.

Throughout the summer the wether and I rehearse the slaughter. At first he panics and screams like a baby when I restrain him, but after a couple of tries he knows the routine and will quickly acquiesce. He is never fully tamed, but once lying down with his horns and ears on the ground and his jugular exposed to the sky, he nearly falls asleep. He'll let me stroke his neck, rub his belly, talk to him, and he'll remain there a while after I step back from him leaving him, breathing softly, not even twitching his tail.

Shortly after the last sassafras leaf falls to the ground, the wether and I rehearse this routine for the last time. His carcass is hung from the garage rafters for the afternoon while I watch the first snowflakes of the year fall to the ground. The offal, which filled less than half a five-gallon bucket, is buried in the compost pile where it will disappear in a couple of weeks. I assume a fox dug up the skull because I found it half buried beneath a pine tree while on a quiet walk the day after Christmas. It is always so quiet the day after Christmas, and the sight of that horned skull peeping out from the ground was pure silence.

The carcass is butchered in the garage. The flesh cools quickly and my fingers go numb before I'm halfway through. It smells sweet, and like blood and meat, and like goat. Everything is cut down to meal-size subprimals, wrapped, and frozen. A few large cuts are kept for large dinners: a whole leg, a couple of thick steaks, a set of six saddle chops. The whistling swans coodle-hoo overhead. They fly so high and yet their voices carry so well. It's getting cold.

The long and large bones are all set aside for bone broth. The stubborn pieces of meat and fat are left on. They're piled up on a large pan, doused with sea salt, pepper, and nothing else because I'm in a hurry—a child is screaming. Sometimes I add some tomato paste, thyme, bay, or whatnot. They're roasted until sizzling, a little browned but not burned, then the whole pan is added to a large pot of rolling boiling water, then quickly brought down to a steaming simmer—never boiling. For a good lesson on demi-glace, see the fourth season of Northern Exposure, "The Big Feast." This here is broth, no demi-glace, but do not overboil the broth and be sure to treat it like liquid gold. These bones were well grown and tended to. Let simmer as long as you wish—the longer, the more concentrated and richer the flavor. The house smells like the essence of goat: sassafras, grass, maple, oak, burdock, raspberry, blackberry, rose, pine, all the flowers of the field, honeysuckle, creeping ivy, poison-ivy, kudzu, tiger lily, iris, half my garden, including corn, beans, sorghum, pepper, sage, thyme, parsley, garlic, clover, rye, buckwheat, apple, peach, and pawpaw (some say goats won't eat pawpaw, but mine sure do), and so much else. The broth is done once the marrow is all out of the bones and the bones are demineralized and dull to the eye. Finally, bring it right up to a boil, then strain into clean hot mason jars and can.

WINTER

The garden goes to sleep under straw, a thick sod of winter rye, then a blanket of snow on top. The neighbor's cows won't be seen again until April. Our goats and chickens hunker down, bracing themselves. Everything falls cold and quiet. Nights are long.

It is late January—cold and at times miserable, but the sun is setting a little later and the sunrises are more brilliant. A half bone-on, half boneless goat shoulder is taken out of the freezer, rubbed in sage and sea salt, and marinated in the spring beer.

Today the ground is a sparkling snow-white blanket under a cloudless deep space-blue sky. I start a bread dough and set it aside to rise. My three-year-old daughter and I roll the dough stuck to our hands in little balls and throw them at each other. Outside, the sun is bright but no warmth is afforded. We thumb our noses at the cold, bundle up, and start a fire with the oak in the pit outside. Cardinals and chickadees peck at the bird feeder while I curse at the cold stealing the flame from my matches and heat from my fingers. But soon the kids are roasting marshmallows in their snow suits.

Once the fire is roaring and hot, I go back inside and take out four or five large onions from the cellar, slice into rings, place in a large heavy pot and fry in salt, butter, and olive oil on medium heat. Once the onions are slightly browned on the edges, I open a bottle of the spring beer and dump it onto the onions, bringing up any fond. Two quarts of the canned bone broth are added to the pot along with some of the sage. Rub the sage into the pot, and whatever you do, do not wash your hands when finished. The winter gloves I wore that day will smell of sage for a long time.

I take the goat shoulder, which has been marinating in the beer and sage for the last couple of days, out to the fire and smoke it for a couple of hours while I drink the beer and eat burned marshmallows. Snow is tossed onto the fire and meat in order to smoke up the fire and keep the meat from cooking too fast. My wife listens to our muffled play and laughter while napping inside. The girls and I sample some of the meat, which is brown and crispy on the outside and rare in the middle. It's good. "Leave some for the soup," I tell them as I take another bite. And if you, the reader, attempt to duplicate this process, good luck, but eat as much as you wish at this point—you're the cook, and it is really good. The shoulder is smoked and roasted to a golden brown, then put in the pot. Like the broth, the soup is cooked at a steaming simmer, never a hard boil, for a good while until gravity removes all the tender meat from the bone and the marrow is cooked out. The soup slowly darkens to a radiant amber-brown. The richness of the broth will temper the bitter of the beer; the smoke and sage are pure warmth. The bread is taken out of the oven. Sea salt, pepper, sage, and beer are added to taste. Serve the soup piping hot with the fresh rolls. Gather and sit down. Find a comfortable place and ruminate.

EMBRACING DISTURBANCE AS PART OF YOUR SYSTEM

JASON DETZEL

I am a grass farmer managing the disturbance and rest cycles of my pastures through animal pressure. I have made many mistakes along the way, and each season brings new events that test my ability to adjust and recover. What I cannot do is predict what the weather will bring each year. The more time I spend seeing how my soils on my land are affected by the weather, the easier it gets to make the appropriate adjustments. The mistakes I have made are easier to recover from knowing that the system is built upon massive resources and cycles that are constantly evolving. I manage through a holistic lens, but I am also a business and need to balance my management techniques with my income needs. This type of management has allowed me to be an excellent steward of the land while making money and adjusting my impact options based on the season and the needs the plants and animals. Being a good rancher means paying attention to detail and to destruction.

We, as movers of soils and animals, understand that we can manage but cannot lord over nature. The recorded history of our planet is scant and scattered, making our understanding of earth's cycles tenuous at best. And yet we still attempt to control our land, to nudge our properties, and to work against nature when we feel like it is working against our particular enterprise. Every year I get this way around the spring thaw. The weather warms, the water melts, and certain areas of my farm are transformed into a muddy goulash complete with bits of floating detritus and manure. Managing cattle in a holistic system requires attention to detail and attention to destruction. Every year I fret and moan over the loss of production due to the intense animal pressure, but every year I also welcome the verdant sward that is created after the disturbance, and more importantly, after the rest has occurred.

Disturbance is part nature's cycle. Earth's plants and animals have evolved a symbiotic relationship with disturbance and rest cycles, and it is our responsibility to utilize these cyclical occurrences to the best of our properties. Fires, floods, wind events, large mammal tracking, and insect feedings are all considered large-scale disturbances that temporarily change the landscape in which they occur. There are both positive and negative disturbance events, defined by their overall impact on that particular ecosystem. Obviously, systems change over time, but if we consider the system for what it is at this moment, these disturbances can have drastic effects on the land.

Positive events are those that are ultimately productive to the system. In the case of fire, certain tree species are triggered to produce more seeds, the understory is cleared for the process of regeneration, and the system evolves. In the case of the cattle farm, the increased impact and pugging that accompanies heavy hoof traffic in the wet spring conditions will help to bring new seeds to the surface, spread manure, and encourage new plants to compete for spots in the nutrient-rich pasture. The key to the positive impacts of disturbance is rest. For the system to evolve, there needs to be time for the evolving units to express themselves in the new system. In the case of the ranch, these areas are the ones that I will not graze until the very end of the cycle. By managing both the disturbance and the rest, the system is able to recover and evolve to be more varied and productive.

Negative disturbance is most common in monocultures. Nature is built on diversity, and the degradation of a monoculture is a direct result

of the natural world exploiting the vulnerabilities of an unnatural system. No rest can also create negative disturbance patterns. If I were to begin grazing the muddy area of the farm along with the other paddocks, it would further stress the area, leading to decreased production and exponentially increasing the rest time required

Jason Detzel

to restore balance. We as movers of land and animals can learn to understand and embrace the disturbance even if we cannot visualize the patterns and connections that are at its core.

Jason Detzel is the farmer and owner of Diamond Hills Farm, a pasture-based cow-calf operation in Hudson, New York, and is the Livestock Educator for the Ulster County Cooperative Extension Office.

HEURESIS

JILL HAMMER

Heuresis is the Greek word for mother-daughter reunion. It describes the meeting of Demeter, goddess of the earth, and her daughter Persephone, when Persephone returns from a long winter in the underworld.

Oweynagat is a cave in County Roscommon, Ireland. Lore holds that it is the residence of a war and fertility goddess, the Morrigan, and the entrance to the underworld.

The cave floor seeps black mud.
Halfway down the tunnel
is the birthing-stone, and beyond the stone
whatever exists before birth.

This is mother-welcome:
covered by dark and dirt,
Persephone running into Demeter's arms
reversed. The underworld draws me in
like a child into a lap,
as if there has never been a search,
as if reunion has no need for words.

I have eaten every seed in the pomegranate.
I forget I was ever born
and handed across a hallway in a blanket.
Now I am the daughter of rocks
that neither abandon nor forget.

The first light to enter my eyes
took my mother from me,
and when I crawl from the cave
the first light takes her from me again.
Mud on my body
dries in the sun like blood on babies.

From Jill Hammer, *The Book of Earth and Other Mysteries* (Cincinnati: Dimus Parrhesia Press, 2016). Reprinted with permission.

The Vagrants. Walker

MAKING BIOCHAR TO IMPROVE SOIL

BARBARA PLEASANT

Ben Short

By making biochar from brush and other hard-to-compost organic material, you can improve soil—it enhances nutrient availability and also enables soil to retain nutrients longer.

One method of making biochar: pile up woody debris in a shallow pit in a garden bed; burn the brush until the smoke thins; damp down the fire with a one-inch soil covering; let the brush smolder until it is charred; put the fire out. The leftover charcoal will improve soil by improving nutrient availability and retention.

Last year I committed one of the great sins of gardening: I let weeds go to seed. Cleaning up in fall, I faced down a ton of seed-bearing foxtail, burdock, and crabgrass. Sure, I could compost it hot to steam the weed seeds to death, but instead I decided to try something different. I dug a ditch, added the weeds and lots of woody prunings, and burned it, thus making biochar. It was my new way to improve soil—except the technique is at least three thousand years old.

What's biochar? Basically, it's organic matter that is burned slowly, with a restricted flow of oxygen. The fire is then stopped when the material reaches the charcoal stage. Unlike tiny tidbits of ash, coarse lumps of charcoal are full of crevices and holes, which help them serve as life rafts to soil microorganisms. The carbon compounds in charcoal form loose chemical bonds with soluble plant nutrients so they are not as readily washed away by rain and irrigation. Biochar alone added to poor soil has little benefit to plants, but when used in combination with compost and organic fertilizers, it can dramatically improve plant growth while helping retain nutrients in the soil.

AMAZONIAN DARK EARTHS

The idea of biochar comes from the Amazonian rain forests of Brazil, where a civilization thrived for two thousand years, from about 500 BCE until Spanish and Portuguese explorers introduced devastating European diseases in the mid-1500s. Using only their hands, sticks, and stone axes, indigenous Amazonian people grew cassava, corn, and numerous tree fruits in soil made rich with compost, mulch, and smoldered plant matter.

Amazingly, these "dark earths" persist today as a testament to an ancient soil-building method you can use in your garden. Scientists disagree on whether the soils were created on purpose in order to grow more food, or if they were an accidental byproduct of the biochar and compost generated in day-to-day village life along the banks of the earth's biggest river. However they came to be, there is no doubt that Amazonian dark earths, often called *terra preta*, hold plant nutrients, including nitrogen, phosphorous, calcium, and magnesium, much more efficiently than unimproved soil. Even after five hundred years of tropical temperatures and rainfall that averages eighty inches a year, the dark earths remain remarkably fertile.

Scientists around the world are working in labs and field trial plots to better understand how biochar works and to unravel the many mysteries of *terra preta*. At Cornell University in Ithaca, New York, microbiologists have discovered bacteria in *terra preta* soils that are similar to strains that are active in hot compost piles. Overall populations of fungi and bacteria are high in *terra preta* soils too, but the presence of abundant carbon makes the microorganisms

115

live and reproduce at a slowed pace. The result is a reduction in the turnover rate of organic matter in the soil, so composts and other soil-enriching forms of organic matter last longer.

In field trials with corn, rice, and many other crops, biochar has increased productivity by making nutrients already present in the soil better available to plants. Results are especially dramatic when biochar is added to good soil that contains ample minerals and plant nutrients. Research continues, as described by the International Biochar Initiative (www .biochar-international.org), but at this point it appears that biochar gives both organic matter and microorganisms in organically enriched soil enhanced staying power. Digging in nuggets of biochar—or adding them to compost as it is set aside to cure—can slow the leaching away of nutrients and help organically enriched soil retain nutrients for decades rather than for a couple of seasons.

FINDING FREE BIOCHAR

Biochar's soil-building talents may change the way you clean your woodstove. In addition to gathering ashes (and keeping them in a dry metal can until you're ready to use them as a phosphorus-rich soil amendment, applied in light dustings), make a habit of gathering the charred remains of logs. Take them to your garden, give them a good smack with the back of a shovel, and you have biochar.

If you live close to a campground, you may have access to an unlimited supply of garden-worthy biochar from the remains of partially burned campfires. The small fires burned in chimneys often produce biochar too, so you may

need to look no farther than your neighbor's deck for a steady supply.

Charcoal briquettes used in grilling are probably not a good choice. Those designed to light fast often include paraffin or other hydrocarbon solvents that have no place in an organic garden. Plain charred weeds, wood, or cow pies are better materials for using this promising soil-building technique based on ancient gardening wisdom.

HOW TO MAKE BIOCHAR

To make biochar right in your garden, start by digging a trench in a bed. (Use a fork to loosen the soil in the bottom of the trench and you'll get the added benefits of this "double-digging" technique.) Then pile brush into the trench and light it. You want to have a fire that starts out hot but is quickly slowed down by reducing the oxygen supply. The best way to tell what's going on in a biochar fire is to watch the smoke. The white smoke, produced early on, is mostly water vapor. As the smoke turns yellow, resins and sugars in the material are being burned. When the smoke thins and turns grayish blue, dampen down the fire by covering it with about an inch of soil to reduce the air supply, and leave it to smolder. Then, after the organic matter has smoldered into charcoal chunks, use water to put out the fire. Another option would be to make charcoal from wood scraps in metal barrels.

I'm part of the Smokey-the-Bear generation, raised on phrases like "learn not to burn," so it took me a while to warm up to the idea of using semi-open burning as a soil-building technique. Unrestrained open burning releases 95 percent or more of the carbon in the wood, weeds, or whatever else that goes up in smoke. However, low-temperature controlled burning to create biochar, called pyrolysis, retains much more carbon—about 50 percent—in the initial burning phase. Carbon release is cut even more when the biochar becomes part of the soil, where it may reduce the production of greenhouse gases, including methane and nitrous oxide. This charcoal releases its carbon ten to one hundred times more slowly than rotting organic matter. As long as it is done correctly, controlled charring of weeds, pruned limbs, and other hard-to-compost forms of organic matter, and then using the biochar as a soil or compost amendment, can result in a zero-emission carbon cycling system.

Burning responsibly requires simple common sense. Check with your local fire department to make sure you have any necessary permits, wait as long as you must to get damp windless weather, and monitor the fire until it's dead.

THE BIGGER PICTURE

If global warming is to be slowed, we must find ways to reduce the loss of carbon into the atmosphere. In the dark earths of the Amazon, and in million-year-old charcoal deposits beneath the Pacific Ocean, charcoal has proven its ability to bring carbon release almost to a standstill. If each of one million farmers around the globe incorporated biochar into 160 acres of land, the amount of carbon locked away in the earth's soil would increase fivefold.

But there's more. What if you generate energy by burning a renewable biomass crop (like wood, corn, peanut hulls, bamboo, willow, or whatever), while also producing biochar that is then stashed away by using it as a soil amendment? The carbon recovery numbers in such a system make it the only biomass model found thus far that can produce energy without a net release of carbon. Research teams around the world are scrambling to work out the details of these elegantly earth-based systems.

Much remains to be known about how biochar systems should tick, but some may be as simple as on-farm setups that transform manure and other wastes into nuggets of black carbon that help fertilizer go farther while holding carbon in the soil.

As gardeners, it is up to us to find ways to adapt this new knowledge to the needs of our land. To make the most of my bonfire of weeds, I staged the burn in a trench dug in my garden, and then used the excavated soil to smother the fire. A layer of biochar now rests buried in the soil. Hundreds of years from now, it will still be holding carbon while energizing the soil food web. This simple melding of soil and fire, first discovered by ancient people in the Amazon, may be a "new" key to feeding ourselves while restoring the health of our planet.

To learn more about this fascinating topic, read Johannes Lehmann, *Amazonian Dark Earths* (Boston: Kluwer Academic, 2003).

Originally published as "Make Biochar—This Ancient Technique Will Improve Your Soil," *Mother Earth News*, February–March 2009, www.motherearthnews .com/organic-gardening/making-biochar-improve-soil -zmaz09fmzraw.

YARROW:

Guardian of Our Psychic Boundaries

LAINE V. SHIPLEY

In the beginning of my herbal medicine journey, yarrow was the first plant I connected with. She's powerful, grounded, and protecting. She thrives in harsh environments that concentrate her medicine, making it stronger. She is often found on the borders of humans and the wild, growing on the edges of winding highways. When well watered, she grows bushy and soft. In desert climates, her single stalks grow straight and spare with her bone-white flowers and umbeled crown upon her head. Her roots grow in large interconnected mats just below the soil's surface.[1] Wafting through the meadow breeze is her pungent aroma—sweet and spicy, a scent that always reminds me of the first time I saw her.

Yarrow, or *Achillea millefolium*, is a powerful healer of physical and emotional ailments. After Achilles was born, his mother held him by the heel and dipped him a bath of yarrow to protect him from harm. With this protection he became a legendary warrior, until he was wounded in the heel she held.[2] *Achillea millefolium* translates

Laine Shipley

120

to "one-thousand-leaved herb of Achilles"[3] and became Achilles's ally in battle. He treated his soldiers' wounds with yarrow, using the plant to staunch the bleeding.[4] The leaves and flowers are a potent hemostatic, and the fresh or dried plant can be used as a poultice on bleeding wounds. Dropping a bit of tincture or oil on a gauze can aid in staunching the bleeding of a deep cut.[5] Yarrow works especially well on cuts to the bone. You can see an indication for this in the shape of her serrated leaves, which are cut back to the rib.[6]

Internally, yarrow is useful for treating menstrual cramps and can help regulate periods. If your period is heavy, yarrow will help ease the flow, and if your period is light, yarrow can increase the blood flow. Her anti-inflammatory actions help relax the uterus and alleviate cramping. Yarrow should not be used during pregnancy.[7]

Aside from her connections to blood, yarrow is known as a folk medicine for treating fever and flu. At the onset of a fever, drink yarrow tea and draw a hot bath. Yarrow brings the blood to the surface of the body, raising your body temperature, which causing sweating.[8] This helps break the fever to allow your body's healing processes to start.

My favorite aspect of yarrow's healing comes in the psychic realm. Sitting in a field of yarrow, I become grounded. I can feel her moving from my center to my crown, in warm energy. I envision her growing inside me, her flower crown radiating above my head. She shines her light down in a protective shower dropping down from her umbel flowers. Yarrow has long been a mystical herb connected to witchcraft and divination.[9] In Chinese medicine, yarrow stalks are used in casting the I Ching. Wise women healers of Europe used yarrow to treat female ailments. Much of their medicinal knowledge has been lost due to the persecution of witches in the Middle Ages,[10] but yarrow still has folkloric use in prophetic dreams, love spells, and "dark brews."[11]

Yarrow protects psychic boundaries and is extremely helpful with people who are easily affected by other's emotions.[12] She is known as the Herb of the Wounded Healer because she will teach you how to heal yourself.[13] This is an important boundary to protect in modern times. We are constantly exposed to emotions and energies from strangers that deplete our energetic reserves. One drop of tincture on the tongue or a lightly brewed cup of tea will help. If yarrow is not handy, I call to her and invoke her spirit. She is always with me, and always protects me. Use yarrow to help you manage in any occupation that put you in frequent contact with people.[14] She will protect your boundaries so you can decide for yourself how much you want to give to others. She reminds us that in order to be healers, we must heal ourselves first. In order to take care of others, you must take care of yourself.

NOTES

1. Michael Moore, *Medicinal Plants of the Pacific West* (Santa Fe, NM: Red Crane Books, 2001), 272–75.
2. Rosemary Gladstar, *Herbal Healing for Women* (New York: Fireside, 1993) 259–60.
3. Laura C. Martin, Wildflower Folklore (Charlotte, NC: East Woods Press, 1984), 142–44.
4. Gladstar, *Herbal Healing*.
5. Scoot Kloos, "Introduction to Psycho-Spiritual Plant Medicine," lecture for Elderberry School of Botanical Medicine, Portland, OR, February 28, 2013.
6. Matthew Wood, *The Book of Herbal Wisdom: Using Plants as Medicines* (Berkeley, CA: North Atlantic Books, 1997), 64–83.
7. Gladstar, *Herbal Healing*.
8. Moore, *Medicinal Plants*.
9. Martin, *Wildflower Folklore*.
10. Gladstar, *Herbal Healing*.
11. Moore, *Medicinal Plants*.
12. Kloos, "Psycho-Spiritual Plant Medicine."
13. Wood, *Book of Herbal Wisdom*.
14. Kloos, "Psycho-Spiritual Plant Medicine."

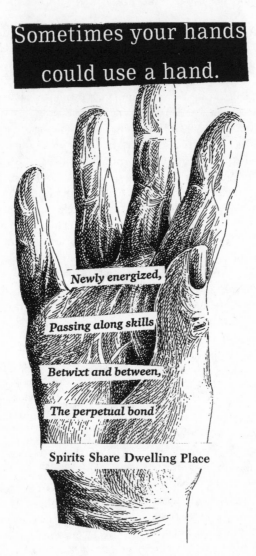

Sometimes your hands could use a hand.

Newly energized,

Passing along skills

Betwixt and between,

The perpetual bond

Spirits Share Dwelling Place

Rachel Weaver

GREEN SMILES, PASTURE BREATH, AND TELESCOPIC VISION

STEVE SPRINKEL

You're getting closer. Closer to the record volumes of vegetation consumed by mid-evolutionary Homo sapiens eighty thousand years ago. These graze-worthy masticators chewed through four to six pounds of leafage a day. Keep at it. The munching energy alone was dietetic. They simply had to keep eating in order to keep eating. Early wise hominids made their own sugar, cellularly, without the benefit of photosynthesis, in the minute vortices of the liver. Such vast consumption eventually led to a self-induced green-matter intoxication, which is one reason why some of you feel nearly as addicted to kale as an alcoholic is to vodka. Without the hangover. And the more vegetation the sapiens consumed, the smarter they got. They purportedly ate beet greens so often that with the naked eye they could see the rings on Saturn and invented an early form of astrology.

Early green geniuses suddenly discovered in 77,516 BCE that to intentionally produce desirable roughage was preferable to wandering about hoping to run into a nice field of spinach untrammeled by antelope and tapirs. The ability to make a choice was the result of gobbling these brainy foods. These opportunities to decide between one prospect or another no doubt created giddy pleasure among our forbearers. Imagine the instantaneous glee they experienced when surveying riversides stocked with bamboo, artichokes, and collards.

"I'll start with the collards, with pressed mint sauce, and have the artichoke leaves for my entrée. The bamboo is one of my favorites, but isn't it a little out of season?"

"Yes, madam. You are right. My mistake. We actually stopped serving the bamboo earlier this month. I gave you an old menu in error. Instead, we are now featuring White Nile watercress on a bed of early malva."

"Is the malva still free range? In the foraging grounds on the Euphrates they have started to hand-select the varietals and, call me a purist, but I really think they have sacrificed flavor for palatability. Not to forget the spectacular phyto-nutrition found in the wild. Gnawing all day on raw long-fiber cellulose is what first had us standing erect, after all, and if we still hope to one day invent Ska, the Nintendo, and the AC-130 Ghostrider Ground Support Gunship, I don't think we should trifle with non-heirloom roughage. Do you? We'll need all the freshly sourced potherbs we can incessantly stuff if the species is going to achieve it's gloried destiny someday, don't you think?"

"Of course, madam. That is why chef Verdant provides you the option of edible table service. The crunchy, hand-pressed Yangtze River bokchoy serving plates go well with the watercress."

All of this easy living, creekside with a wad of cud in one's maw, gave rise, shall we say, to resultant procreative urges, and as the population grew, culinary horizons expanded as well. Anthro-paleontologists suggest this is why you find hominid carnivores who depend more and more on animal flesh increasing in greater numbers as one travels away from the equator until one is an Inuit and all that's edible are seals and unwary cetaceans.

The Circumpolar Inuit, Laplanders, Nenets, Enets, Khanty, and Yukaghir peoples may have in fact enjoyed eating something green other than lichens. Recent scientific discoveries utilizing carbon-14 analysis indicate that during intermittent thaws, receding ice exposed fertile ground near their once-frosty villages, where migratory birds dropped seeds of edible plants brought from southern climates. Fragments of plants vaguely related to daikon, long bok, and gai lan were found encased in ice. Chopsticks and small bottles marked "soy sauce" also were discovered nearby.

Editor's note: Steve Sprinkle owns Farmer and the Cook in Ojai, California, with his wife, Olivia. He grows vegetables and was a longtime contributor to *Acres* magazine. He serves on the board of the Cornucopia Institute and Eco Farm and is a major benefactor of young farmers in the greater Santa Barbara area. You can subscribe to Steve's musing "The Forager," and we highly recommend it. —SvTF

APRIL

IMPROVISATIONAL AND URBAN COMMONS

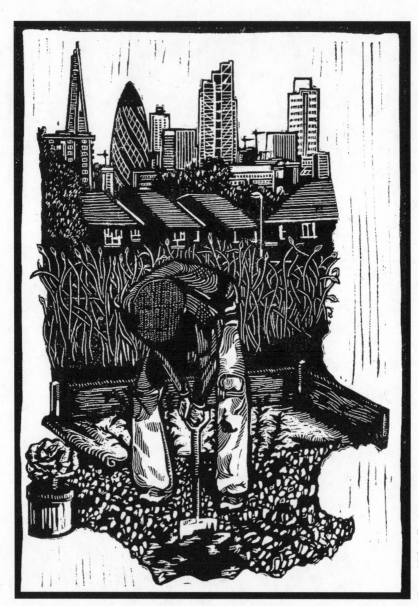

Queen Mob Collective

YOU ARE A FOX

SARAH CHANDLER

You are a fox
Your stripes are familiar
Shadows block light and allow angles of focus

I am chilled by the wind
I am warmed by your glow

I want to know your hiding places
To follow your scent

Is this the shrub after all?
I notice some buds, not yet showing their color

Perhaps I am looking too closely
What else did the red tail hawk have in view?

PSILOCYBIN THERAPY:

Mycroremediation for the Psyche

T. CODY SWIFT

For most Americans, the modern world offers the unwavering impression of our triumph over the natural world. Despite looming climate change, devastating storms, and historic drought in California, most people's daily experience tells us little of our susceptibility to the dynamics and rules of the natural world. For the most part, fresh produce is consistently and abundantly available regardless of season or weather, our lives are not threatened by animal predators or parasites, and infectious disease has virtually disappeared relative to developing countries. But despite our vast accomplishments against the pressures of nature in these areas of health and agriculture, the United States has more anxiety and mental illness than any other culture and than any other time in history, with nearly a quarter of our citizens regularly taking psychiatric medications. We must ask ourselves: Has our technological departure from the natural world contributed to this looming sense of despair, confusion, and isolation? Do plants and fungi, especially those with psychedelic properties, have the potential to remind us of our interconnection with the earth and its cyclical processes? And, perhaps most

importantly, can these medicines offer some remediation for not only our mental health but the health of an ailing planet?

In the recent psilocybin trials at Johns Hopkins and New York University, especially in those for cancer patients with anxiety, there may be some insight into how plant sacraments may indeed help to heal our broken psychological connection with nature. Having a life-threatening illness like cancer opens the door of awareness to our mortal existence and fundamental vulnerability to nature, leading to debilitating psychological suffering as patients try, and often fail, to reconcile this new reality. For nearly a decade, patients with cancer-related anxiety have had the opportunity to have a single high-dose session with psilocybin, the active compound in magic mushrooms, in a supportive setting with trained guides. During the psilocybin session, many participants describe powerful visionary experiences where they are afforded a deep feeling of interconnection with the world and others. From this place of new awareness, they often gain profound confidence in the face of death, lasting well beyond the single session. Below are quotes from three different participants describing their experience in the psilocybin trial. The first is from a woman with ovarian cancer who came into the study having been told she had a 50 percent chance of living: "You realize you're going to die. I don't know that you realize that until you're told." During her session she had a transformative vision where she saw herself underneath the forest floor:

I'm in the forest and there's this beautiful, loamy, woodsy, green lush kind of woods, and I'm down below the ground … and it felt really, really good, and I thought, "That's what happens when you die." I am going to be reconnected with this beautiful world. This earthy world that we live in … and it was just simple. It was gorgeous.

[The psilocybin] just opens you up and it connects you.… It's not just people—it's animals, it's trees—everything is interwoven, and that's a big relief.… I think it does help you accept death because you don't feel alone, you don't feel like you're going to, I don't know, go off into nothingness. That's the number one thing—you're just not alone.

I feel like a whole bunch of crap has been dumped off the surface. This stuff that made my world shut down so much and made me look at the ground and watch the clock numbers clicking by. There's life and so many things going on, just watching that tree over there blowing in the breeze, seeing people in the street, and all the different people in vehicles rushing by! I just feel good about being alive.… It's always there, we just don't notice, and I'm trying to notice and not forget that I can see it at any time, I can hear it any time.

Following these experiences, participants in the study reported significant reductions in anxiety and depression, as well as a greatly reduced fear of cancer progression and recurrence. They described being much more at peace in the present moment, and slowing down to savor the subtleties of their social, emotional, and natural surroundings. It seems the psilocybin provides a deep-felt reminder of the vibrancy and cyclical nature of life, where death is no longer a fearful exile but an inclusive part of that whole. When asked how this psilocybin treatment may help others, one participant said that it brings "people to a level of awareness that fear and anxiety don't have the nourishment to grow from. That seems to be how it works."

Consciousness has always allowed humans a certain degree of distance from the natural world through thought, language, and other ways we represent the world. Rituals with psychoactive plants have been used for millennia, with psilocybin mushrooms among the early Indian Vedic cultures and the Mazatec people of Mexico, ayahuasca in the Amazonian basin, and peyote cactus in Mesoamerica. These ancient traditions afforded transcendence from these human tendencies, and not only reaffirmed but also truly celebrated our interconnection with the natural world. Although technology has taken us farther from the earth in many ways, it may now be helping to bring us in contact with the exact tools that hold the potential to reconnect us— with traditional medicines being brought north from Central and South America, and through the synthesis of naturally occurring psychedelic

compounds. Over the past century, and especially the past decade, these plant substances have dramatically increased their political and geographic reach, showing up in Food and Drug Administration-approved research studies and in the great proliferation of traditional ceremonies across the United States and Europe. Just as mushrooms show up to breakdown organic materials that are waiting to be returned to earth, these psychoactive plants are emerging at a time of great disconnection, anxiety, and ecological crisis. These plant medicines are helping to deconstruct our own sense of separation, and remind us, in humbling, beautiful, and awe-inspiring ways, the preciousness of where we came from, and to where we shall return.

T. Cody Swift, MA, is a counselor, qualitative researcher, and board member of the Heffter Research Institute, the primary sponsor of current trials with psilocybin. To learn more, visit: www .heffter.org.

THE BEEKEEPER'S CLOCK

SUSANNAH BRUCE HORNSBY

*For to the bee a flower is a fountain of life and to
the flower a bee is a messenger of love.*
—Kahlil Gibran

I live in in the slate-velvet crook of the Blue
Ridge Mountains of Virginia, in an old crookedy
farmhouse that was once the main house for
two hundred stately acres of grazing land and
orchards. Generation after generation of its
previous owners passed on without wills, and
so, piece by piece, like ham off a prize pig, most
of the land was lost before we got our hands on
it. We're left without enough land to farm and,
frankly, deep-set in its rockbed, run through
with steep-sided ravines and seams of glittering
quartz, our land doesn't lend itself to the plow.
Instead, this red-dirt beauty offers her richness
with riots of wild blooming flowers, flowers
much visited by bees.

As dawn breaks on a warm day, the new
light touches down on the domain of the bees.
This is their forage, and over this they preside.
All of the open blossoms and kiting currents of
air that carry the deep and timeless secrets of
nectar and possibility and information into the
hive on the wings of far-flung honey-makers,
all of the freshwater springs, every wildflower
median, teeming byway, and fourth-generation
orchard becomes their quarry, their commons.

It's a symbiosis, it is a collective, it is a simple,
natural dominion of highest-and-best-use: the
miraculous, mysterious, dizzyingly complex bee
systems, evolved over multiple millennia, the
good labor of bees being bees.

As soon as the frost breaks in our part of
Virginia, the forsythia is first to bloom, followed
closely by the flowering quinces. The bees don't
work crocus, really, nor daffodils so much,
or even peonies when the time comes; but in
spring's first tentative flush, you'll see the pale
gray pollen of the red maple and the hot yellow of
the alder, and then the tulip poplar trees, which
my grandfather called "junk trees" and which
you wouldn't even notice were blooming unless
you parked under one by accident and your car
got a confectioner's sugar dusting of its palest
golden pollen and covered in its petal-soft tulips.
If you're not a beekeeper or a bee loyalist, you'd
probably say "shit!" quietly and forcefully under
your breath. If you are a beekeeper in Virginia,
you can see the bees streaming in and out of
the hive, their baskets bulging with the creamy
pale yellow bolls for weeks of intermediate
sustenance before the major heat and sticky-
sweet fertility and buzzing frenzy of summer
and its ripe-garden glories are spread out before
us like a hot breakfast. For those who care to
look up, the tulip poplar is a glory, and the major

source of nectar and pollen throughout the pre-spring chill.

Though it is a largely solitary and meditative practice, beekeeping attracts a certain sort of person with its concurrent threads of logic, magic, and beauty: the scientist and the maverick, the naturalist, the storyteller, and the shaman. The bee yard stays full of mysteries, loveliness, and a constant stream of small tragedies. We are members of the Central Virginia Beekeeping Association, and one of my great, old-fashioned pleasures is to sit around and listen to our fellow beekeepers shoot the bull. I listen to them talk about the seasons, this winter compared to last winter, who lost how many hives, their plans accordingly. New gear or methods they had tried, what worked, what didn't, bears, bad queens, good queens, the weather, always the weather, and, owing to the weather, what's in bloom.

Hearing the old beekeepers talk about what's blooming versus what should be blooming puts me in mind of the Linnean Flower Clock. Dreamed up by Swedish botanist Carl Linnaeus, who also devised the Latin naming system for plants, the Horologium Florae was a formal

garden in the shape of a clock, the plantings based on the notion that certain flowers open and close at the same time every day, and that, therefore, if you planted the varietals in a circle, you could tell the time by the sequential blooming and unblooming of the fragrant blossoms.

Just like the morning glory and the night-blooming cereus are tuned and wound up somehow deep in their mystic pistils to bloom like clockwork, so is the seasonality of the tulip poplar and the hawthorn, of henbit and dandelions and goldenrod and tiny fall asters, and all of the lesser ditch-blooms that my bees unapologetically prefer over showy, nectar-poor town flowers. The steady unfolding and marking of time, carefully calibrated by tender blossoms in an endless unbroken sequence of budding, opening, quickening, and fluttering down. All perfectly timed to coincide with the bees' visitations, of course.

This has been an exceptionally warm December. The forsythia already exploded in a riot of yellow and the quinces are blushing their deep naughty pinks. The hellebores are practically freaking out, and my friend's apple orchard is covered

in buds, as are my tulip poplars. Those darling bees do their duty, and take their pleasure like ladies out on the town on their small, weak, winter wings, feeling the flush of hive-freedom that usually doesn't come until March. Their industry is awesome and gorgeous, so I watch them. They work each individual calyx in descending order of richness, a meritocracy of sweet yield. They pay me no mind, barefoot in my sleeveless dress on the winter solstice—the nectar flow is on, sister.

But what will happen when the frosts return? These delicate blossoms and their fresh, sweet nectars are meant for spring. Nature's hour strikes too soon, the clock is unwound, the flowers bloom backward, the springs are sprung, the mechanism unreliable. The bees have been betrayed in their expectations. The covenant of our shared universe seems to hang in the perilous balance of some new, dark clockwork—tuned by unseen hands to some unknown end. I watch them at their work and can't help but feeling that I, and my wild blooming
valley, have failed them
somehow.

How well the skilful gardener drew
Of flow'rs and herbs this dial new;
Where from above the milder sun
Does through a fragrant zodiac run;
And, as it works, th' industrious bee
Computes its time as well as we.
How could such sweet and wholesome hours
Be reckoned but with herbs and flow'rs!
—Andrew Marvell, "The Garden"

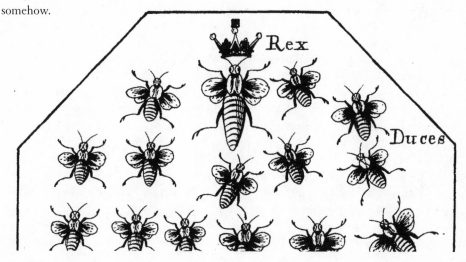

TRANSFORMATION:

A Call to Action

CORY W. WHITNEY

This September, at the Terra Madre meeting in Torino, Italy, University of California, Berkeley, professor and agroecology activist Miguel Altieri challenged the Slow Food movement and the mission of "Good, Clean, and Fair" food for all. He argued that the movement has to find a way to make sure that poor and vulnerable people access such food. This is a story of the movement making such access possible. It is the story of the radical transformation of food systems and indigenous self-determination in development through tactical commoning. It is the story of communal sanctuary gardens of traditional knowledge by indigenous Bakiga farmers in southwest Uganda.

Many long years of colonialism followed by guerrilla fighting and civil war destroyed much of Uganda's traditional food systems and food traditions. Despite recent political stability, the people of Uganda still struggle in systems of corruption, poverty, severe food insecurity, and hunger. The loss of traditional home gardens with the commensurate loss of traditional crops, horticultural practices, and food systems has contributed to the severe food insecurity of the region.

In Uganda's southwest, recently displaced indigenous Bakiga farmers can be found living on the edge of the Kashohe-Kitome rain forest, where they still maintain their traditional home gardens. These dynamic and complex traditional food systems help them to sustain their communities in the face of extreme vulnerability. Poverty, systemic corruption, disease, and food insecurity lead to high rates of child mortality and lower life expectancy and quality.

The Bakiga traditional food system is based on each family maintaining a subsistence-based home garden. These diverse intercropped systems of bananas and other fruits are designed to meet all family nutritional and medicinal needs. Any money that can be earned, such as from selling the odd sack of beans or bunch of bananas, is used to pay school fees and perhaps to buy some salt or soap, which are not all that common in households.

Slow Food's 10,000 Gardens in Africa Project was started as an answer to the problem

John Craxton

of food insecurity in Africa. As Slow Food vice president Edie Mukiibi said at Terra Madre this year, "Even though people say we are hungry, we still need to eat 'Good Clean, and Fair' food." Slow Food president Carlo Petrini said, "I would like to ask the missionaries to stay home. I have a lot of confidence in Africans. Africans know how to do things for themselves." The overarching idea is that local African farming communities can organize and engage in developing their own systems of food security and food sovereignty through interaction with the Slow Food community.

The 10,000 Gardens in Africa Project in this Bakiga community is an example of tactical commons. Through the gardens, the communities have decided that traditional knowledge can be preserved in the common space. The struggle against the systems of poverty and destruction of traditional knowledge happens collectively and openly as a solidarity movement. The gardens serve as a sanctuary for traditional crops, traditional farming knowledge, and traditional horticultural practices. Slowly, these community gardens are coming together with the enthusiasm and energy of the youth and the traditional knowledge of the elders as well as the financial support of the Slow Food convivia in Germany.

One big question for us, as Slow Food activists, was about how to get it started. This was a carefully orchestrated approach generating cooperation among everyone involved. We had to consider how to present the idea of a communal garden in the village and how to present Slow Food and the radical notion that the community members, already very busy with maintaining their own gardens, finding school fees, dealing with illness and extreme poverty, now should take on the task of maintaining their dying Bakiga food and horticultural traditions. In a community where each family produces its own food in a small home garden, building gardens in the commons requires the generation of a new social architecture. The enthusiasm and active voice of the youth together with the support of the village chair were all part of the establishment of the garden.

Without the support of Slow Food Uganda and financial support from the German convivia, the project would not be possible. Slow Food Uganda has helped bring the message of Slow Food as a solidarity movement to the hearts of the community members. Knowing that they are not the only people facing these problems, and knowing that they have the support of other poor communities and peasant farmers around the world and across Uganda, has inspired the communities to action. German Slow Food members have supported the project by sending funds to the 10,000 Gardens in Africa Project as well as bundles of gardening tools. The convivium members in Germany even had the forethought to send rechargeable solar lights and headlamps, knowing that the twelve hours of darkness at the equator makes it hard to maintain a garden.

Now the Bakiga people work together to grow a common garden. The orchestration of the establishment of this project was complicated and slow by definition. For the Bakiga communities on the edge of the Kashohe-Kitome rain forest, the Slow Food 10,000 Gardens in Africa Project is part of the global resistance against corporate ownership and capitalist usurpation of traditional knowledge and indigenous peoples' lands. In this case, the Slow Food garden is a sanctuary for traditional crops and traditional knowledge.

It is critical that establishing the garden begins with a dialogue with local communities about the purpose of the garden and the idea behind Slow Food. The garden should serve community needs while also stopping the loss of traditional knowledge and the loss of traditional crops and crop practices. If organizers are not clear about the intention of the Slow Food movement, communities may choose to plant economic gardens for generating funds for the community. Herein lies a danger for the Slow Food movement in implementing these gardens. Another danger is that the gardens will be lost over time without sustained activist solidarity and financial support from Slow Food. If the gardens are to be sustainable, recognition and funding from the international convivia should be regular and sustained. Convivia abroad should agree to adopt Slow Food gardens and pay regular fees for their upkeep and for further development.

Editor's note: Learn more about what's going on in Africa at the Oakland Institute, which focuses on land grabbing and other development scandals that displace traditional peoples from farm-based livelihoods with false promises, corruption, and debt traps. It's cool to think about Slow Food offering a platform for more horizontal engagement between agricultural communities in the North and South. —SvTF

THE HOP PROJECT

The Hop Project is a contemporary art project that will tour Herefordshire, Worcestershire, Birmingham, and the Black Country in 2016–17. Funded by Arts Council England's Strategic Touring Programme, the project uses the historical migratory movements of hop pickers as the conceptual basis for a touring exhibition. The Hop Project is conceived and produced by General Public (artists Elizabeth Rowe and Chris Poolman).

The project's starting point is an exploration of the social and political implications of hop production in the West Midlands. Hops are the flowers of the hop plant *Humulus lupulus*. They are used primarily as a flavoring agent when brewing beer, into which they impart a bitter, tangy flavor. In the West Midlands, the hop yards of Herefordshire and Worcestershire produce more than half of the hops grown in the UK.

Historically, in the nineteenth and early twentieth century, a mass exodus from Birmingham and the Black Country used to occur every autumn as thousands of people traveled to Herefordshire and Worcestershire for the hop picking season. The Gypsy, Roma, and Traveler communities also have a long history of participating in agricultural work and were a significant source of flexible, short-term labor in the hop picking industry at this time. This all ended abruptly in the 1960s when mechanization of the hop industry brought an end to the need for large numbers of workers to support the annual harvest. Since this time, dramatic changes have taken place in agriculture and in the workplace more generally. One of the outcomes of these developments is that today Herefordshire has an ever increasing Central European community traveling to the county to fulfill agricultural demands.

The touring exhibition includes:

> Archival material relating to the history of hops and hop picking. This includes a television report from 1966 in which presenter Lionel Hampden interviews some of the last Black Country hop-pickers in Herefordshire.

> The Hip Hop Pickers' Ball—an amalgamation of hip-hop and hop picking. The exhibition features album cover designs and a music video featuring a young rapper

from Birmingham called Tee Lyrical. This video charts the journey undertaken by hop-pickers between Birmingham and Herefordshire.

> Contributions from groups situated along the tour's geographical route. These include watercolors inspired by inner city graffiti (Cradley, Mathon, and Storridge Art Group), quilted beer label designs for fictional feminist ales (Crystal Quilters and Kidderminster Forest Quilters) and a model of the final "Hop Pickers' Special" train to leave Herefordshire (West Bromwich Model Railway Society).

Other works in the exhibition include a map documenting the annual movements around the West Midlands undertaken by a Romany Traveler family, photos of "agricultural sculpture" in Herefordshire, a Great Western railway map from the 1930s, and a film by Polish artist Alicja Rogalska (Broniów Song), a contemporary folk song based on conversations with local people in the Polish village of Broniów.

HOP, HOP, HOPPING

The project doesn't seek to present a factual, social history of hops; rather it offers an interpretation from a number of different angles and perspectives. It uses the verb "hopping" as a working methodology to explore a number of tangential ideas connected to the history of hop growing. The phenomenon of economic nomadism is a particularly important idea in the wider project. Indeed, an exploration of the historical and contemporary movement of people for the purposes of work is highly appropriate for a touring "moving" exhibition.

Movements of other types underpin the project. These include white van movements, Romany Travelers' figure of eight movements, and the mapping of premiership footballers' birthplaces. Other ideas the project is interested in include the politics of geographical boundaries, representations/myths of the countryside, employment rights, the demise of railways, the building of motorways, the effects of mechanization within agriculture, itinerant lifestyles, and the relationship between rural and urban tribes with their different "languages" and modes of communication.

THE TOUR

The Hop Project will tour for eighteen months across the West Midlands. The tour schedule corresponds to both the historical movements of hop pickers and hop growing areas. During this time it will visit fifteen different venues including The Courtyard Centre for the Arts (Hereford) from January 13 to March 19, 2017 and Mac Birmingham from September 9 to November 12, 2017.

For more information about the Hop Project, visit http://thehopproject.co.uk.

EAT YOUR SIDEWALK!

SPURSE

Eat Your Sidewalk! is just what it says. We believe that it is time to change everything about how we eat, think about food, and engage with our urban ecosystems. We believe it is time to start foraging and eating our sidewalks. Change needs to begin right where we are. Foraging the weeds in the cracks of our streets right under our feet, and not in some far off pristine forest, is a delicious, joyous activity that has the capacity to spark deep and far-reaching ecological change. When you bend down a pick a dandelion growing from a crack in the street, what has happened to this plant now happens to you—your fates are joined. You are of this place in a way you have never been. This is a profound act with important consequences for us, these weeds, our eating habits, and our sense of place.

Eating your sidewalk is about the complex and entangled ways of being alive. This extends beyond us to include the very active participation of plants, animals, histories, technologies, ideas, other worlds and practices. For far too long we have held ourselves apart from these worlds—socially, politically, economically, and ecologically. We wish to change this in a direct, engaged, and joyous manner.

To help foster a sense of being of a place that could open us up to the wonders of meeting radically distinct worlds and remake our own world we have developed a program we call Eat Your Sidewalk! It is one part foraging and one part commons building and all centered on the social and intraspecies pleasure of eating.

What does "Eat Your Sidewalk!" mean? For us it is in those three simple words:

Eat: Eating is what links us to all life. Our health and its health are linked. Its concerns and ours meet. We can no longer separate our fates.

Your: You are not alone as you pick this plant. Others (both human and non-human) also want it: you have to negotiate and work together. This means forming a community based on shared concerns (health, sustainability, pleasure). Not a top-down community, or a community in name only, but a co-evolving community of partners.

Sidewalk: So often we talk about local but skip over our actual place to get to the parts of our environment we more easily recognize because they are more like products or have been defined for us as important. But this means we are not fully addressing our actual environment. How do we do this? Begin with where you are— your sidewalks, yards, neighborhoods, and the systems that they are part of—and pay attention to everything. When this really happens a place comes alive.

!: Lets not forget the thin bit of punctuation! The exclamation mark is there because it has to be as urgent as it is fun. (It can't be all doom and gloom). Eat Your Sidewalk! celebrates the excesses that just might come with a less dependent way of life.

We are very excited and pleased to announce the completion of our Eat Your Sidewalk! cookbook. It includes exciting recipes that run the gamut from the wonderfully weird to the extremely pragmatic; personal stories from those on the frontline in rethinking food and ecology; deep discussion and analysis of the entangled history of ecological relations; and philosophical speculation on the power of urban foraging, If you are interested in purchasing a copy of the cookbook, send us an email at inquiry@spurse.org. For more information on Eat Your Sidewalk!, visit www.spurse.org/what-weve-done/eat-your-sidewalk/.

EAT YOUR SIDEWALK
PROCEDURES TO MAKE THINGS COMMON

WHY EAT YOUR SIDEWALK?

To develop new models for our urban lives and local ecosystems – but these won't fit within our traditional politics framed along a LEFT-RIGHT continuum with its false either/or choices, & "middle" grounds.

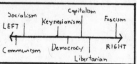

Our major political options occupy a field that's organized by 2 axis: INDIVIDUALITY-COLLECTIVITY & FREEDOM-HIERARCHY

These political models miss what's critical & pleasurable about eating our sidewalks. This exists at the far end of another axis: HAVING – DOING. Eating your sidewalk convivially locates itself at the far end of the doing spectrum (while contemporary politics stresses having). Doing happens in the middle of things with others – many others.
This common middle is our focus.

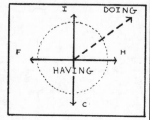

A GLOSSARY

HAVING

Our everyday lives are primarily focused on three categories of seemingly discreet things: resources, individuals & needs. This shapes our engagement with the local into questions of having (ownership & rights). Thus "being local" happens primarily through buying. Solutions appear as things to have: commodities, legislation, or state sponsored services. This is part of an individualistic entrepreneur-consumer outlook, which often fosters decreased cooperation, trust & accountability in all of us.

DOING

Eating your sidewalk begins with the realization that we are actively of the world (doing). We're always already engaged with dynamic fields and processes. As you fuse your health with that of dandelions, you become focused on making and unmaking patterns, networks, assemblages and communities. In moving from having "the local" to a Worldly doing we actively make things common. Sidewalk Eating makes processes common to all – we're becoming consciously <u>Intra-active</u> <u>& intra-dependent</u> with our urban ecosystems.

EAT YOUR SIDEWALK – THE PROCEDURES
So how do we begin? Start in daily life sensing how you are al[...]

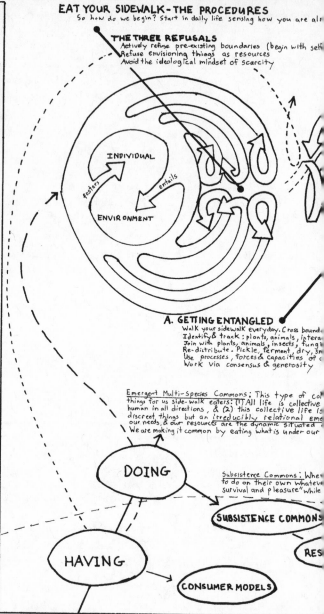

THE THREE REFUSALS
Actively refuse pre-existing boundaries (begin with self[...]
Refuse envisioning things as resources
Avoid the ideological mindset of scarcity

A. GETTING ENTANGLED
Walk your sidewalk everyday. Cross bound[...]
Identify & track: plants, animals, intera[...]
Join with plants, animals, insects, fung[...]
Re-distribute. Pickle, ferment, dry, sm[...]
Use processes, forces & capacities of c[...]
Work via consensus & generosity

<u>Emergent Multi-Species Commons</u>: This type of co[...]
things for us side-walk eaters: (1) All life is collective[...]
human in all directions, & (2) this collective life is[...]
discreet things but an <u>irreducibly relational</u> eme[...]
our needs, & our resources are the dynamic situated[...]
We are making it common by eating what is under our [...]

<u>Subsistence Commons</u>: Whe[...]
to do on their own whateve[...]
survival and pleasure" while [...]

COMMONS

There is a long history of local communities whose resisted the enclosures of our <u>individualistic</u> <u>consumer-entrepreneurial</u> <u>paridigm</u>. This resistance, paired with a vision of reality as inherantly collective, is the commons (or commoning). Eating your sidewalk joins these wonderful ongoing practices.

EMERGENCE

This process is key to sidewal[...]
(systems) are not merely the [...]
they are <u>irreducibly relation</u>[...]
independent status (sensing [...]
to commoning). Emergent syst[...]
their internal components in [...]
non-linear manners. Our s[...]
remake us.

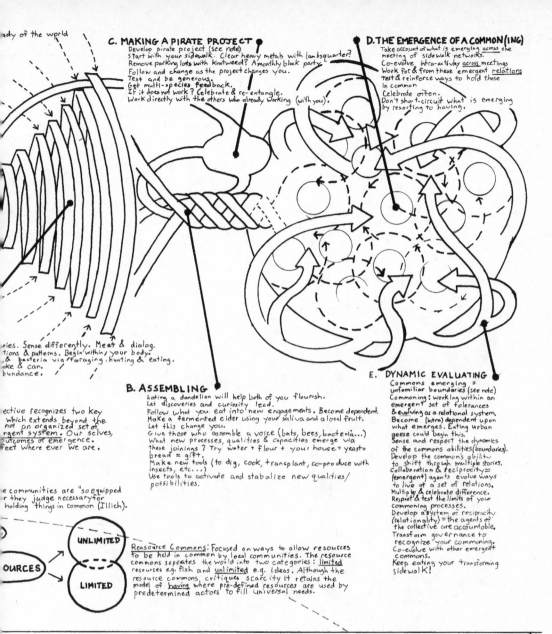

ady of the world

C. MAKING A PIRATE PROJECT
Develop pirate project (see note)
Start with your sidewalk. Clear heavy metals with lambsquarter?
Remove parking lots with Knotweed? A monthly block party?
Follow and change as the project changes you.
Test and be generous.
Get multi-species feedback.
If it does not work? Celebrate & re-entangle.
Work directly with the others who already working (with you).

D. THE EMERGENCE OF A COMMON(ING)
Take account of what is emerging across the
meeting of sidewalk networks.
Co-evolve intra-activity across meetings.
Work for & from these emergent relations.
Test & reinforce ways to hold these
in common.
Celebrate often.
Don't short-circuit what is emerging
by resorting to having.

ries. Sense differently. Meet & dialog.
tions & patterns. Begin within your body.
& bacteria via foraging, hunting & eating.
oke & can.
bundance.

ective recognizes two key
which extends beyond the
not an organized set of
rgent system. Our selves,
utcomes of emergence.
feet where ever we are.

e communities are "so equipped
r they judge necessary for
holding things in common (Illich).

B. ASSEMBLING
eating a dandelion will help both of you flourish.
Let discoveries and curiosity lead.
Follow what you eat into new engagements. Become dependent.
Make a fermented cider using your saliva and a local fruit.
Let this change you.
Give those who assemble a voice (bats, bees, bacteria...)
What new processes, qualities & capacities emerge via
these joinings? Try water + flour + your house = yeast =
bread = gift.
Make new tools (to dig, cook, transplant, co-produce with
insects, etc...)
Use tools to activate and stabalize new qualities/
possibilities.

E. DYNAMIC EVALUATING
Commons emerging =
unfamiliar boundaries (see note)
Commoning: working within an
emergent set of tolerances
& evolving as a relational system.
Become (intra) dependent upon
what emerges. Eating urban
geese could begin this.
Sense and respect the dynamics
of the commons abilities (boundaries).
Develop the common's ability
to shift through multiple stories.
Collaboration & reciprocity =
(emergent) agents evolve ways
to live of a set of relations.
Multiply & celebrate difference.
Respect & test the limits of your
commoning processes.
Develop a system of reciprocity
(relationality) = the agents of
the collective are accountable.
Transform governance to
recognize your commoning.
Co-evolve with other emergent
commons.
Keep eating your transforming
sidewalk!

UNLIMITED

OURCES

LIMITED

Resource Commons: Focused on ways to allow resources
to be held in common by local communities. The resource
commons seperates the world into two categories: limited
resources e.g. fish and unlimited e.g. ideas. Although the
resource commons, critiques scarcity it retains the
model of having where pre-defined resources are used by
predetermined actors to fill universal needs.

BOUNDARIES

A boundary in an emergent commons is not
where something stops or is made to stop
(as is the case with enclosures). This is why it
is so hard to talk about "the local". Boundaries
are where something begins to emerge. The
relational emergence of boundaries develop
with a system's capabilities. These relations
are sensed as qualities, capacities, and processes
and should be considered the basic processes
of commoning. They are what is held in common
as your eating evolves with your street.

PIRATE PROJECT

Walking & eating your streets generates great ideas
and that is the beginning of a Pirate Project.
Its goal is two-fold: to pleasurably entangle
you deeper into a world & to catalyze a new
emergent process. We call it "pirate" and not
"pilot" because it's not about being scalable or
universally applicable. It seeds many distinct
initiatives which form a network as they evolve.
What emerges across this network is what is most
critical (this is a commons in the making).

k eating. Things
sum of their parts,
, & have no
these relations is key
tems transform/make
dynamic and
sidewalks will

SPURSE

MAY
DIGITAL

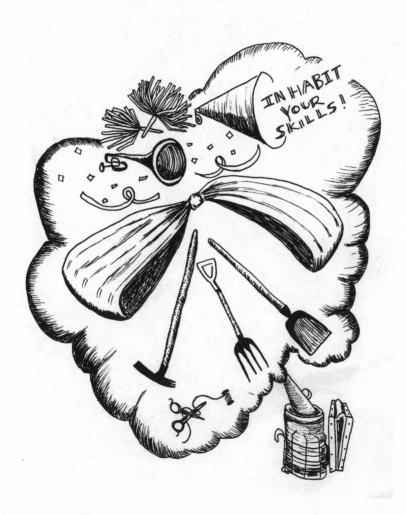

R. de Salis

GATHA

JEREMY HARRIS

My condition reflects the theme of 21st-century life. Good enough.
It's neither thriving nor ailing, neither exciting nor dull.
I'm both content and in eternal want of more.
Not rich, not poor, still somewhat ambitious, already exhausted.
I never work, yet I never have a day off.
Activity is constant, most of it anticlimactic.
I try to transcend profound indifference.
Perhaps this and other successes are near.

TOOLS

BINYAMIN KLEMPNER

Yoni Gantcher

As a small-scale, landless, wandering urban farmer, small hand tools are my most important and reliable asset. Even if I get booted off the land I've been farming on for a season or two, my tools are mine to take with me. As such, I have developed a system of farming that works well for me based on the tools I have and on my need to travel light and make the most of the limited growing space available to me. These are the seven tools I need more than any other to work the land, presented in the order I use them:

1. Grub hoe: Lightweight, easy to use, this classic tool is my one-stop shop for breaking new ground, weeding, hilling, chopping, and initial bed preparation.

2. Broadfork: Deep tillage without creating a hard-pan and without harming the soil structure and the life that lives within the soil. I have both a Creature Meadows and A Valley Oak broadfork; I decide which one to use depending on soil type.

3. Soil crumbler: After broadforking, I take my Wolf-Garten soil crumbler and lightly till the top two inches of soil.

4. Soil rake: Get that bed smoothed and ready for seeding.

5. Trowel: Great for transplanting.

6. Dail-a-Seed seeder: Made by Wolf-Garten, this sturdy tool clips onto any Wolf-Garten handle. It takes up little space in my backpack and works very well.

7. Small three-tined cultivator: This is my tool of choice for aerating, weeding, and reworking soil in between crops. I use the Wolf-Garten brand as it is sturdy and fits into my backpack. I clip it onto a Wolf-Garten handle when I'm ready to use it.

There are a bunch more tools that are nice to have on hand to make the job easier, like a dibbler, Johnny's Easy-Plant jab-type planter, and Hoss's garden seeder and steel-wheel hoe, but for me they're unnecessary. My seven-tool system utilizes tools that make my work compact, systematic, and transportable in an urban setting—qualities required by all poor urban farmers lacking long-term affordable land access.

A PERSONAL REFLECTION ON MOBILE PHONES AND GMOS

DANJO PALUSKA

> I watered the cult earlier. Ran the irrigation
>
> Thanks?
>
> They get cranky waiting for the ufos when they're dry
>
> Read 4:40 PM
>
> Up top. Turned it off a while ago
>
> Cakes. Cunts. Cukes. Fuck you, Autocorrect!
>
> Love you mom
>
> Delivered

I was a high-tech human, living an adolescent dream placed in my head by television and reinforced by too much grad school. I was making robotic artificial limbs, robot art, and practicing "Internet, open source, shiny happy people with soldering irons fixing the world." For a variety of reasons it was ceasing to add up. Then I fell down a rabbit hole realizing I had never really made anything from scratch. (All high technology is built on slave labor and a toxic hole in the ground somewhere else.) Nothing grounded. Concrete floors and pixel eyes, starving for land.

With some luck and help from others, I find my way to being a farmworker for a couple years, then on to primitive skills and off-grid for a bit after that. I went in starry-eyed, looking for some magical Amish wonderland free of the technology I had stuffed myself with, but I was still in America. I found cell phones and computers at every turn, and people asking me about robots to help on farms. The physicality was wonderful. Moving things around outside and feeding people really is fantastic. The financial realities were frightening. Eventually I find myself back to working in technology part-time for money and working land part-time for life. Still today I'm processing this reality and thinking about how digital technology is often viewed as OK but GMOs are not. To me, these are sides of the same coin.

So here is a little compare-and-contrast with two friends from my journey, mobile phones and GMOs. Both of these characters are certainly in

149

my life, although I would like to have a little less of each. They are both everywhere. Some eighty percent of all cotton, soy, corn, and beets (sugar) are GMOs, and when is the last time you met someone without a cell phone? I do know two people over age sixty who don't have them, but they both have a partner with one. Can you walk down a street in a city without seeing someone staring at a screen? Or someone with a GMO-sugared drink in their hand? Or wearing cotton clothing?

Independent of anything else about the health of individuals, society, and the environment, these are the products of very large corporations. They're not coming from a Shaker village. So is it strange that a large number of organic farmers have their mobile phones as a key part of running their business? Or that people who choose a bucolic rural life would choose to farm GMO corn? Or that all the urbanites who buy organic food have a large percentage of their daily communications through digital and inorganic matter? This is just life in the twenty-first century. Here are some comparisons between the two:

> GMOs are designed to encourage farmers to spray poisons (dollar cost) on their crops, and these end up in the soil, water, and animals. Cell phones encourage all sort of content (mental bandwidth cost) propagation that can be likened to poison for local social ecosystems.

> GMO seeds are not owned by farmer customers, so they cannot save seed and replant next year as they would with other crops. Cell phones are unmaintainable, throw-away-and-upgrade devices, and you have to buy everything from the company app store.

> Both are produced by large corporations with big CEO and executive paychecks.

> GMOs create genetic connections that mother nature would not likely come to on her own. These might do strange things to local ecosystems or our bodies. Mobile phones create all sorts of social connections across time and space, very unnatural in the animal and tribal sense of bonding, communication, and community.

> GMOs don't encourage farmers to be truly in touch with their land. Key resources (seeds, fertilizer, and pesticide) come from off-farm. Mobile phones encourage people to pay more attention to national news items, and even for local news items, messages are sent off to remote servers before being passed to a local person.

It's a short story and a short list, but consider what makes it true to you. Then use your organic communication devices and let me know what you think.

Danjo Paluska
Postcard Aficionado
12 Dunning St.
Brunswick, ME 04011

WHICH ARE YOU?

—

There are two kinds of people on earth today,
Just two kinds of people, no more, I say.
Not the sinner and saint, for 'tis well understood
The good are half-bad, and the bad are half-good.

Not the rich and the poor, for to count a man's wealth
You must first know the state of his conscience and health.
Not the humble and proud, for in life's span
Who puts on vain airs is not counted a man.

Not the happy and sad, for the swift flying years
Bring each man his laughter, and each man his tears.
No: the two kinds of people on earth I mean
Are the people who lift, and the people who lean.

Wherever you go, you will find the world's masses
Are always divided in just these two classes.
And oddly enough, you will find, too, I ween,
There is only one lifter to twenty who lean.

In which class are you? Are you easing the load
Of overtaxed lifters who toil down the road?
Or are you a leaner, who lets others bear
Your portion of labor and worry and care?

—Ella Wheeler Wilcox.

THE BLACK LIST

OLIVIA TINCANI

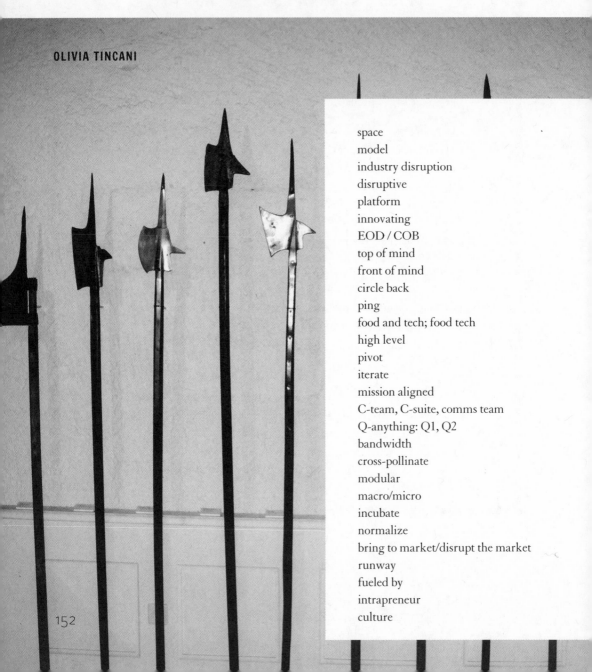

space
model
industry disruption
disruptive
platform
innovating
EOD / COB
top of mind
front of mind
circle back
ping
food and tech; food tech
high level
pivot
iterate
mission aligned
C-team, C-suite, comms team
Q-anything: Q1, Q2
bandwidth
cross-pollinate
modular
macro/micro
incubate
normalize
bring to market/disrupt the market
runway
fueled by
intrapreneur
culture

These are the words created, co-opted, coerced out of the layperson's lexicon and into the Valley vernacular currently spewing out from just south of a city I once thought of as home, San Francisco. This is the historic birthplace of the freedom of speech, civil rights, farmworker rights, and honest diversity.

These words live in the playbook of a burgeoning brood of start-ups, CEOs, founders, funders, investors, early adopters, and future unicorns that have made the city their pitch. These words no longer hold fast to their original meanings. That the bees and the butterflies have been toppled from their airy, aimless flight and pollinate no longer has much to do with preserving biodiversity, or sex. Culture, somehow, has tumbled and tripped into this new trap as well: the beloved better half of agriculture has been reappropriated to signify the modern-day water cooler conversation. Culture is the millennial's version of what was once just bad holiday punch and brown carpet: company off-site team-building trips to Burning Man, on-demand subscription-based smoothies, entire conference rooms dedicated to major motion picture themes, and a cacophony of media surrounding you while you "work." This all now means culture, the expression of our core elements, the microecology of ourselves within the context of our surroundings.

In the new lexicon, people incubate ideas of convenience instead of eggs. We fuel fund-raising rounds instead of small engines. Macro and micro refer to the size of silicon-based bits and bobbles instead of the mushy insides of our cellular structures. And we bring it all to the stock market, an original permutation of the agricultural merchant's market, the spice trader's market, the livestock market, the produce market, the commodities market. Debt was born inside agricultural trading mechanisms, and the word holds the weight of its provenance of integrity.

The inhabitants of the technology space have borrowed much of the agrarian's language, perhaps without asking, or perhaps without us much even noticing. How could we let our words go so easily? How do we hold them back from reappropriation? Or rather, should we? Is it theft or a nod to their root, source, and inspiration? Should we embrace the fluid flex of language as a signifier of opportunity and harbinger of both the future and the past? Perhaps we should just adapt, as one does in nature, in farming, in chaos, and in technology-based, venture capitalist–funded high-growth companies, and thus get on with thriving.

153

PEASANT WEAPONS

Bardiche: a type of two-handed battle ax known in the sixteenth and seventeenth centuries in Eastern Europe

Bill: similar to a halberd but with a hooked blade form

Fauchard: a curved blade atop a six-foot, seven-inch pole, used in Europe between the eleventh and fourteenth centuries

Ge (戈): a Chinese weapon in use from the Shang Dynasty (1500 BCE) that had a dagger-shaped blade mounted perpendicular to a spearhead

Glaive: a large blade, up to eighteen inches long, on the end of a six-foot, seven-inch pole

Guandao (关刀): a Chinese polearm from the third century CE that had a heavy curved blade with a spike at the back

Guisarme: a medieval bladed weapon on the end of a long pole; later designs implemented a small reverse spike on the back of the blade.

Ji (戟): a Chinese polearm combining a spear and dagger-ax

Kamayari (鎌槍): a Japanese spear with blade offshoots

Lochaber ax: a Scottish weapon with a heavy blade attached to a pole in a similar fashion to a voulge

Naginata (薙刀): a Japanese weapon with a twelve- to twenty-four-inch-long blade attached by a sword guard to a wooden shaft

Partisan: a large double-bladed spearhead mounted on a long shaft with protrusions on either side for parrying sword thrusts

Pole-ax: an ax or hammer mounted on a long shaft, developed in the fourteenth century to breach the plate armor increasingly being worn by European men-at-arms

Ranseur: a pole weapon consisting of a spear-tip affixed with a cross hilt at its base, derived from the earlier spetum

Spontoon: a seventeenth-century weapon consisting of a large blade with two side blades mounted on a six-foot, seven-inch pole; considered a more elaborate pike

Voulge: a crude single-edged blade bound to a wooden shaft

War scythe: an improvised weapon that consisted of a blade from a scythe attached vertically to a shaft

Welsh hook: similar to a halberd and thought to originate from a forest-bill

Woldo (월도): A Korean polearm with a crescent-shaped blade mounted on a long shaft, similar in construction to the guandao

From Wikipedia, https://en.wikipedia.org/wiki/Halberd#Similar_and_related_polearms.

HELMETS

TOUCHING ENCOUNTERS:

A New, Integrative Commons

BEATRICE VERMEIR

The Tree of Life

The emergence of a notion of the commons was born in Europe in the sixteenth and seventeenth centuries when, for the first time, the meadows and forests that previously held grazing or foraging rights became subject to mass privatization. The delineation of common grounds from private property that occurred as a result was undoubtedly an important moment in which the definition and management of commonly accessible natural resources was refined. In the following century the Age of Enlightenment, which was dominated by philosophies of rational individualism, reinforced the tone of what was to be a trajectory of increasing privatization for territories the world over—what Foucault called the urge to "order" what is seen to be an unruly and chaotic nature—that reaches right up until the present day.

As this trajectory runs its divisive course, our understanding of the commons becomes ever more expansive; in the 1980s, the conflict over common ground reemerged with renewed vigor as natural resources—fossil fuels, soil, and water—as well as social structures, such as education and healthcare, began to be sold off by neoliberal governments to private enterprise. In their seminal book *Commonwealth*, Hardt and Negri later expanded the discourse of the commons to include not only natural and social resources but those generated by culture: languages, information, emotional and behavioral codes, for example, which are shared and sustained through social interaction and perhaps less easily removed from common ownership. Even more recently, the definition has been expanded further to envelop the growing mass of digital data generated by humans: the so-called "Informational Commons."

But all these conceptions of the commons are limited to include only what Ranciere called the "sensible": those resources invested with human agency and therefore perceived as valuable or useful to us. In a geological era defined by humankind's decimation of nature and long-term alteration of the earth's stratosphere, in which the future can only conceivably be imagined as either post-nature or post-human, there exists an ever more pressing need to attend to the nonhuman forces in play.

What happens to the definition of the commons when we take a vibrant materialist perspective that must not only include resources as filtered through human agency but takes as its focus the elusive currency of vibrancy: a shared quality of potential energy common to all things, dead or alive, organic and inorganic, human and nonhuman. Is it possible to imagine or even enact a commons in which all things are active agents of their own destinies and where points of contact between those agents can materialize only in spontaneous encounters altogether outside of ownership? And, finally, can we employ this utopian vision of the world as a "heterogenous monism of vibrant bodies" to facilitate the integration of the spheres of personal, social, and environmental experience and do something to patch up our frayed relations with the world as a whole.

Beginning by shedding, as far as possible, the humanocentric view of the entire external world as a resource, maybe we can start to imagine this new, all-encompassing commons. The first port of call in transcending a dualistic, oppositional concept of "nature" and "people," and opening ourselves to the possibility of vibrant encounters, is to acknowledge the "other" in ourselves, and in doing so, materializing the relation between the self and the external world. Vibrant encounters occur, after all, at the point of contact between the self and the external world, where that which is known collides with what can be a fearful and indeterminate unknown. In the words of Lee Ufan of the School of Things: "Encounters are entailed in the fact that other things than oneself exist and it is possible to enter into dialogue with them. An encounter is an interaction with externality/otherness."[1]

Human bodies, Merleau Ponty declared, occupy a precarious position, belonging both to "me" (the self) and the external world of things. For centuries, the Western doctrines of Cartesian "mind over matter" philosophy combined with Christian denials of the flesh in favor of the spirit or soul, what Ufan called an "extreme version of the self," have resulted in increased trust and attention placed in the sphere of intellect and a simultaneous alienation of the body from its state as matter. Our bodies, complex and dynamic mixtures of cells, fluids, vibrations, electrical pulses, thoughts, and dreams, are in fact vibrant material systems—not just one but an "array of bodies"[2] that can touch and affect other bodies, both human and nonhuman, and in turn be affected. In this way, we can also begin to subvert the traditional sovereignty of the intellect, reason, and self over the body, material, and other, and, in the words of Jan Favre, "make bodies think, minds feel."[3]

Touching otherness is transcending the self. By leaving the realm of experience, which is dominated by intellect and mediated by sensory perception, we can establish an attunement to moments that, inexplicably by those terms, prick our attention.

NOTES

1. U-hwan Yi, *The Art of Encounter* (London: Lisson Gallery, 2004). Originally written 1973.
2. Jane Bennett, *Vibrant Matter: A Political Ecology of Things* (Durham, NC: Duke University Press, 2010), 21.
3. Quoted in Paul de Bruyne, Pascal Gielen, and Tilde Björfors, *Community Art: The Politics of Trespassing* (Amsterdam: Valiz, 2011).

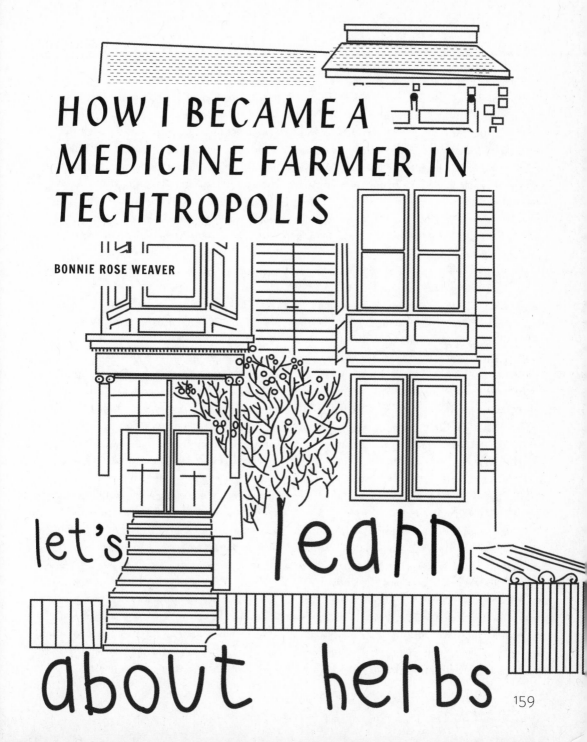

HOW I BECAME A MEDICINE FARMER IN TECHTROPOLIS

BONNIE ROSE WEAVER

let's learn about herbs

There was a unique calmness that came over me when I pushed open the door from the garage into the sunlight of the Guerrero Street garden. The plant spirits greeted me and the bustle of the street melted away. I became entranced by the life of the garden. Thanks to the hard work of those who had tended the space before me, there were figs and loganberries to feast on while I harvested and tended the bounty of medicinal herbs that lived here.

At first I did it for myself; gardening had become my affordable mental health care. But as the garden grew, so did my passion to share it with others. I soon had a very large hobby—or a small business—on my hands. I wanted the project to grow; I wanted to prove to people that medicinal herbs could provide healing in the twenty-first century. I took the idea to city hall and gave it a name: 1849 Medicine Garden.

The year 1849 was when gold was "discovered" in California. That year changed this land in everlasting ways and connects all of us—those who are here now, those who pass through, and those who, over the centuries, have been driven out. This felt like the right name for an herbal medicine farm in San Francisco's digital age; 1849 speaks to our past and present: we live in a gold rush city.

As a child of this city I grew up with asthma. I used steroid and other pharmaceutical inhalants on a daily basis. When I was twenty-one years old, I learned how people used plants to heal themselves and wanted to try it for myself. After some study, I blended an herbal tea to strengthen my lungs and drank it every morning. Within three months I stopped having symptoms of asthma and never again used an inhaler. Herbs

had cured a potentially lifelong affliction that pharmaceuticals never would.

I was empowered and wanted to share this feeling: the feeling of healing one's self with the plants that surround us. To encourage others to connect to land and plants. To listen to each other, to hear our needs in a busy, buzzing city.

There was a lot to learn; growing medicinals is quite different from growing vegetables. But I knew the land was special. When you sat under the fig tree you could almost forget you were steps away from the bustle of Guerrero Street. I felt called to do this work. I began farming, harvesting, tincturing, and selling seed-to-bottle medicine.

Two years into the project, I launched a community-supported agriculture (CSA) program and was well received by San Francisco's urban agriculture and herbal communities—we've sustained over a year of twenty members who receive a tincture and a chapbook every month. The medicine was more than what was in the small brown glass bottles. The garden attracted volunteers eager to learn about herbs, neighbors excited about plants, acupuncturists ecstatic that there was herbal medicine growing in the city. We became a community of people united by the desire to know and explore the medicine of the plants that live in San Francisco and to connect with the earth underneath the city streets.

There is power in this. There is power knowing where your medicine comes from and how it works. There is power in a resilient herbal community. Building this healing community in San Francisco was a political act. It was a peaceful way of saying "no" to pharmaceuticals.

"Yes" to what is available in our bioregion. "Yes" to celebrating our history and to coming together to create a sense of place.

The garden is closed now. We were forced to leave by our landlords in the fall of our third year. Displacement is a common fate for low-income families and community-based projects in our current boomtown. But our herbal community is still intact—I continue to provide monthly medicine through the CSA, I teach classes in my own home and garden. We have grown stronger from the power of the garden; and while it was temporarily uprooted, we have not lost our potential for new growth and regeneration.

passion flower

This food is the gift of the whole universe.
The earth, the sky, and much hard work.
May we live in a way that makes us worthy to receive it.
May we transform our unskillful states of mind, especially our greed.
May we take only foods that nourish us and prevent illness.
We accept this food so that we may realize the path of practice.

—Plum Village Meal Blessing

JUNE

FAITH BASICS

FALLING IN LOVE OUTWARD

DERRICK JENSEN IN CONVERSATION WITH DAVID ABRAM

Derrick Jensen: How would you convince a skeptic that a river, or a mountain, is alive?

David Abram: Actually, Derrick, I'm not interested in convincing anyone that this is true in some objective, literal sense. It seems to me that the literal view of the world is often part of the problem. I'm not trying to get people to just replace one view of what is literally the case with another view of what is absolutely literally true.

I know, however, that we cannot change the way we live, the way we interact with the world, without changing the way we speak. We currently speak about the world in a very goofy way that holds us apart from it, and makes us feel like we're outside, and hence able to control it, master it, manipulate it. There are other ways of speaking that hold us in a very different relation to the world. I don't think that any of these ways of speaking are "true" in some utterly objective sense. I think they're all just different strategies for speaking, different ways of wielding our words. And one strategy, it seems to me, leads us into a richer way of life, into a deeper reciprocity

with the land around us, and with the myriad beings that compose this land.

This is a very different notion of truth from the one that holds sway right now within conventional science, which is still trying to figure out the "truth" of "how nature works." It seems to me that a more fruitful understanding of truth would ask how we can live in right relation with this rainforest, so that neither we nor the rainforest are suffering. If we're going to study humpback whales, how can we as a human community and a humpback whale community flourish as parts of the same world? I'm not interested in pursuing the questions of: What is a humpback whale? How does it work? What are its mechanisms? To even ask those questions presumes that I am something other than an animal myself—that I am some kind of bodiless mind, a pure spectator of nature, rather than a participant in it.

So there's this problem with much of what we've been talking about. Within our contemporary technologized civilization, it is all too easy to say "that rock is alive, that tree is aware and awake." It's too facile, because it's so simple for people to just translate this into their objectified, literal view of the world, and to believe: "Oh, so it is literally alive and aware and awake." It feels to me too much like a perpetuation of our current way of speaking, which uses language to dominate the world, rather than to make contact with the world around us, to touch things, and to feel them touching us, to respond to things. At this strange cultural moment in the West, our way of wielding words is even more of a problem than the content of those words. Of course, when we speak of the world around us merely as a conglomeration of objects, that is a problem. But even more of a problem is that when we speak, we speak as though nothing else is listening. We speak as though none of these other beings can hear what's being said, or can be influenced by our speaking.

DJ: As though nothing else is listening…

DA: Yes. Not in the sense that the birds or the trees could understand the dictionary definitions of our words. I mean, none of the creatures around here—the coyotes and the ravens and the magpies—know the denotative meanings of the words I say, but they can nonetheless hear the tonality in our speaking. They feel the rhythm in our words. They can hear the music and the melody in our conversation. And in that sense some of the meaning comes across. Yet we speak as though nothing else hears, as though we needn't take care how we speak of these other beings. We like to assume that language is a purely human property, our exclusive possession, and that everything else is basically mute.

But what I'm suggesting is that those of us who work to heal or mend the rift between humankind and the more-than-human earth oughta pay more attention to how we speak, we ought to be way more mindful about how we wield our words. If you already know that you're entirely a part of this wild world, if you've already entered, now and then, into a deeply felt reciprocity with another species, or have tasted a profound kinship and solidarity with the living land around you, still: it is not easy, today, to find a way of speaking that does not violate that experience, that does not tear you out of that felt rapport. It's really hard to flow your phrases in a manner that invokes and encourages that reciprocity, or even allows it. Our civilization is masterful at twisting even our most beautiful words to make them into slogans for a commodity-based reality. Our habits of speech have coevolved with a violent relation to the world for so many centuries that one cannot step out of them very easily.

Given the power of this crazed culture to co-opt even the best of our terms, I think that even more important than the content of what we say is the style of our saying, the form of our speaking, the rhythm of our rap. Somehow the music and the texture of our speaking has to carry the meaning, has be appropriate to the meaning at every point. Our deepest intent makes itself felt in the cadence, in the rhythm and the melody of our discourse. If we are not, in fact, disembodied minds hovering outside the world, but are sensitive and sentient animals—bodily beings palpably immersed in the breathing body of the world—then language is first and foremost an expressive thing, the patterned sounds by which our body calls to other bodies, whether to the moon, or to the geese honking overhead, or to another person. It is really a kind of singing, isn't it? Even the most highfalutin and abstract discourse is still a kind of song, a way of singing the world. It may be a really lousy song—a song that's awfully insulting to many of the beings that hear it, one that grates on the ears of owls and makes the coyotes wince—but it's still a song. And those of us who are working to transform things, we're trying to change the tune, to shift a few of the patterns within the language.

In a sense, we all have to become poets. I don't mean that we should be writing poems for poetry anthologies; rather, that our everyday speaking has to touch people bodily as well as mentally. We have to notice the music in our speaking, and take care that the music has a bit of beauty to it, so that we're not just talking as disembodied minds to other abstract minds but as sensuous and sentient creatures addressing other sensuous creatures, so that our animal bodies are stirred and are brought into the conversation, and so that the other animals are not shut out either. We feel their presence nearby, and so we take care not to violate our solidarity with the animals and the animate earth.

DJ: When I write, I don't want anyone to say, "What a great idea." I want them to burst out sobbing, or to become agitated: to have a bodily response.

DA: Uh-huh. When I write, I sometimes feel I'm in service to the life of the language itself. Maybe I write to rejuvenate that life, to open it back on to the wider life of the land, so it can draw sustenance from there. I'm working to return meaning to the more-than-human terrain, which is where all our words are rooted in the first place. I guess I really don't think language, or meaningful speech, is a particularly human thing at all—it seems to me that language is a power of the earth, in which we're lucky to participate.

So I guess for me, then, the question is not really "Is the world alive?" but rather "How is it alive? How does that life hit us? How can we let it sing through us?"

DJ: If traditional indigenous cultures speak of the world as animate and alive, and if, as you've suggested, our own most immediate and spontaneous experience of the world is inherently animistic, disclosing a nature that is all alive, awake, and aware—then how did we ever lose this experience? How did civilized humankind lose this participatory sense of reciprocity with a living world? How did we tear ourselves out of the world?

DA: Lots of factors. Settlement. The development of large-scale agriculture, which entailed fencing out the wild. The emergence of agricultural surpluses, which often led to hierarchical forms of control and distribution of those surpluses. Urbanization. New technologies. But I also believe it had a great deal to do with one of the oldest and most powerful of our technologies: writing. And, in particular, the alphabet.

But in order to understand why, you have to recognize that the animistic experience is not just a sense that everything is alive, but also an awareness that that everything speaks, that everything, at least potentially, is expressive. The evidence suggests that this is baseline for the human organism, an experience common to all our indigenous ancestors. For most of us today, it seems an extraordinary and unusual experience, but in fact it could not be more ordinary and more normal. The normal human way of encountering the world and the things around us is to sense that they are also encountering us, and that they are experiencing each other, and to sense as well that the things are speaking to one

wind going over the leaves.

DA: Sure. Language is just the wind blowing through us.

DJ: I took us down a side alley. You were saying.

DA: That everything speaks. The howl of a

Pieter Bruegel

another, and to us at times—not in words, but in the rustle of the leaves—

DJ: Which are quite possibly tree words. It hit me about a year ago that there is no difference between us speaking and trees speaking. We both use the wind, or maybe the wind uses us. The wind going over the vocal cords and the

wolf, the rhythms of cricket song, but also the splashing speech of waves on the beach. And, of course, as we were both suggesting, the wind in the willows. To indigenous people, there are many different kinds of speech. Many manifold ways of pouring meaning into the world. But if that is our normal human way of experiencing the world, how could we ever have lost it? How could we ever have broken out of that animate, expressive field into this basically mute world that we seem to experience today, where the sun and moon no longer draw salutations from us, but just arc blindly across the sky in determinate trajectories, and so we no longer feel that we have to get up before dawn in order to pray the sun up out of the ground? How did that happen?

I think one of the factors that has been too easily overlooked until now is the amazing influence of writing. All of the genuinely animistic cultures that we know of—whether we talk of the Haida people of the Northwest coast or the Hopi of the Southwest desert, whether we consult the Huaorani of the Amazon Basin, or the Pintupi or Pitjantjara of Australia—these are oral cultures, cultures that have developed and flourished in the absence of any highly formalized writing system. Animistic cultures, in other words, are oral cultures. And so we should wonder: what is it that writing does to our animistic experience of the world?

I would say that writing is itself a new form of animism, a kind of magic in its own right. Writing makes use of the same animistic proclivity that led our oral ancestors to experience the surrounding world as alive, and to feel themselves spoken to by a passing bird or a cloud. To learn to read is to enter into an intense sensorial participation with the letters on the page. I focus my eyes so intently on those written scratches that the letters themselves begin to speak to me. Suddenly, as we say, "I see what it says." The written words "say" something; they speak to me.

Indeed, that's what reading is. We come downstairs in the morning, we open the newspaper, and we focus our eyes on these little bits of ink on the page, and suddenly we hear voices! We feel ourselves addressed, spoken to. We see visions of events unfolding in other times and places. This is magic. It is not so very different from a Hopi elder walking outside the pueblo; he finds his attention drawn by a large rock at the edge of the mesa, focuses his eyes on a patch of lichen spreading on the flank of that rock, and suddenly hears the rock speaking to him. Or a Kayapo woman who, while walking through the forest, notices a spider weaving its intricate web, and as she focuses her eyes on that spider, she abruptly hears herself addressed by the spider. As other animals, plants, and even "inanimate" rivers once spoke to our oral ancestors, so the inert letters on the page now speak to us. This is a form of animism that we take for granted, but it is animism nonetheless—as outrageous as a talking spider.

The difference, of course, is that now it is only our own human marks that speak to us. We have entered into a deeply animistic participation with our own signs, a concentrated interaction that short-circuits the more spontaneous participation between our senses and the sensuous surroundings. Written signs have usurped the expressive power that once resided in the whole of the sensuous landscape: what we do now with the printed letters on the page, our oral ancestors did with aspen leaves, and stones, the tracks of deer and elk and wolf, with the cycling moon and the gathering stormclouds.

Our written signs have tremendous power over us. It's certainly not by coincidence that the word spell has this double meaning: to arrange the letters of a word in the correct order, or to cast a magic spell—because to learn to read and write with this new technology was indeed to learn a new magic, to enter into a profound new world, to cast a kind of spell upon our own senses. Our own written signs now speak so powerfully that they have effectively eclipsed all of the other forms of participation in which we used to engage. And, of course, it is no longer just our written signs, but our TV screens and our computers and our cars that have us in a kind of dazzled trance. The alphabet is the mother of invention, the progenitor of all our Western technologies. It seems we first had to fall under the spell of the alphabet before we could enter into this fever of technological invention.

I don't mean to be getting down on technology, only to say that many of these very complex technologies could only have emerged from the alphabetic mindset. Nor do I mean to be demeaning the alphabet here—I'm a writer, after all. I'm not saying that the alphabet is something bad, not at all. What I'm trying to say is that the alphabet is magic—that it is a very concentrated form of magic, and that like all magics it must be used with real care. When we just take it for granted, when we don't notice its potency, then we tend to fall under its spell.

So while our indigenous ancestors dialogue richly with the surrounding field of nature, consulting with the other animals and the earthly elements as they go about their lives, the emergence of alphabetic writing made it possible for us to begin to dialogue solely with our own signs in isolation from the rest of nature. By short-circuiting the ancestral reciprocity between our sensing bodies and the sensuous flesh of the world, the new participation with our own written signs enabled human language to close in on itself, enabled language to begin to seem our own private possession, and not something born from our encounter with other expressive beings—from the speech of thunder and the rushing rivers. We no longer sense that language was taught to us by the sounds and gestures of the other animals, or by the roar of the wind as it pours through the trees. Language now seems a purely human power, something that unfolds only between humans, or between between us and our own written signs. The rest of the landscape loses its voice; it begins to fall mute. It no longer seems filled with its own manifold meanings, its own expressive power.

I now look out at nature from within my privileged interior sphere of mental subjectivity, but this subjectivity is not shared by the coyotes or the swallows or the salmon. They now seem to inhabit another world—a purely exterior, objective world. They just do their own thing automatically. Creativity, imagination—for so many of us today these seem purely human traits. The mind, we think, is a purely human thing, and it resides inside our individual skulls. You have your mind, and I have my mind; we have this sense that mind is something that is ours—it's no longer a mystery that permeates the landscape. We own it.

DJ: Why can't we engage our own writing and still engage with an animate natural world as well?

DA: We can. The written word didn't necessitate that we break our sensorial participation with other beings; it just makes it possible for us to do so. It doesn't necessitate that the land become superfluous to us, or that we no longer pay much attention to the more-than-human world. But we no longer need to interact with the land in order to recall all the stories that are held in those valleys, we no longer need to encounter coyotes and dialogue with ravens in order to remember all the knowledge originally carried in the old Coyote tales and Raven tales, because now all that knowledge has been written down, preserved on the page. Once the language is carried in books, it no longer needs to be carried by the land, and we no longer need to consult the intelligent earth in order to think clearly ourselves. For the first time we no longer need to speak to the mountains and the wind, or to honor the land's life with prayers and propitiations, because all our ancestrally gathered insights are preserved on the page.

So, the written word was not a sufficient cause of our forgetting, as we philosophers say, but it was a necessary cause, a necessary ingredient in our forgetfulness.

DJ: This reminds me of something John A. Livingston wrote in *The Fallacy of Wildlife Conservation*. He says that once we reduce our input to everything being mediated by humans, we're essentially in an echo chamber, and we begin to hallucinate. We're sensorily deprived, because we're not getting the variety of sensory stimulation we need. His point is that much of our ideology, much of our discourse, is insane, delusional—hallucinations based on the fact that we've put ourselves in solitary confinement.

DA: I think I share a similar intuition, which I might put a little differently. Our senses have coevolved with the whole of the sensuous world, with all these other sentient shapes and forms, all these other styles of life. Our nervous system emerged in reciprocity with all that rich otherness, in relationship and reciprocity with hummingbirds and rivers and frogs, with mountains and rivers, with an animate, living land that spoke to us in a multiplicity of voices. I mean, human intelligence evolved during the countless millennia when we lived as gatherers and hunters, and hence our intelligence evolved in a thoroughly animistic context, wherein every phenomenon that we encountered could draw us into relationship. Yet suddenly we find ourselves cut off from that full range of relationships, born into a world in which none of those other beings are acknowledged as really sentient or aware. We abruptly find ourselves in a world that has been defined as a set of inert or determinate objects and mechanical processes, rather than as a community of animate powers with whom one could enter into relationship. A dynamic or living relationship is simply not possible with an object.

Today the only things you can enter into relationship with are other humans. Yet the human nervous system still needs the nourishment that it once got from being in reciprocity with all these other shapes of sensitivity and sentience. And so we turn toward each other, toward our human friends and our lovers, in hopes of meeting that need. We turn toward our human partners demanding a depth and range of otherness that they cannot possibly provide. Another human cannot possibly provide all of the outrageously diverse and vital nourishment that we once got from being in relationship with dragonflies and swallowtails and stones and lichens and turtles. It's just not possible. We used to carry on personal relationships with the sun and the moon and the stars. To try and get all that now from another person, from another nervous system shaped so much like our own, continually blows apart our relationships. It explodes so many of our marriages because they can't withstand that pressure.

DJ: That reminds me of something I wrote in my book *A Language Older Than Words*: One of the great losses we endure in this prison of our own making is the collapse of intimacy with others, the rending of community, like tearing and retearing a piece of paper until there only remains the tiniest scrap. To place our needs for intimacy and ecstasy—needs like food, water, acceptance—onto only one species, onto only one person, onto only the area of joined genitalia for only the time of intercourse, is to ask quite a lot of our sex.

DA: Indeed. Our intimate relationships become increasingly brittle. We finally turn toward our lover and say, "I really care about you, darlin', but I'm somehow not quite feeling met. I'm just not being met by you in all the ways I feel that I should be." Of course not! Another person cannot possibly meet us in all those ways that we were once engaged by the breathing world. Even a large bunch of human relationships cannot make up for the loss of all that more-than-human otherness, and it is this that makes our human communities intensely brittle and violence-prone. I don't think we have a hoot of a chance of healing our societal ills, or the manifold injustices we inflict on various parts of the human community, without renewing the wild Eros between ourselves and the sensuous surroundings—without "falling in love outward" (in Robinson Jeffers's wise words) with

this earthly cosmos that enfolds us.

As long as we continue to construe the land as little more than a passive backdrop against which our human projects unfold, we'll continue to close ourselves off from the very sustenance that the human community most needs in order to thrive and flourish. As long as we hold ourselves out of relationship with the surrounding earth, we'll be unable to tap the necessary guidance that we need from the old oaks or the elder ponderosa pines that surround our town, from the winds and weather patterns, from the mountains and rivers. Many of these beings live at scales much vaster than our own, and so can offer us some real perspective and a sense of humility. We simply need their wild guidance. Every human community is nested within a more-than-human community of beings. Until we notice this, many of our human relationships will remain exceedingly fragile, and brittle, and we'll keep slamming each other in frustration, busting each other up with bullets and with bombs.

If you really crave a healthy and lasting relationship with your lover, then instill it with a wider affection for the local earth—for the local critters, and plants, and elements. That affection will hold and nourish your relationship, will feed it and enable you and your partner able to be fluid with one another.

From *How Shall I Live My Life?: On Liberating the Earth from Civilization*, edited by Derrick Jensen (Oakland, CA: PM Press, 2008).

CUCURBITUS Iᴱᴿ

Costume d'apparat.

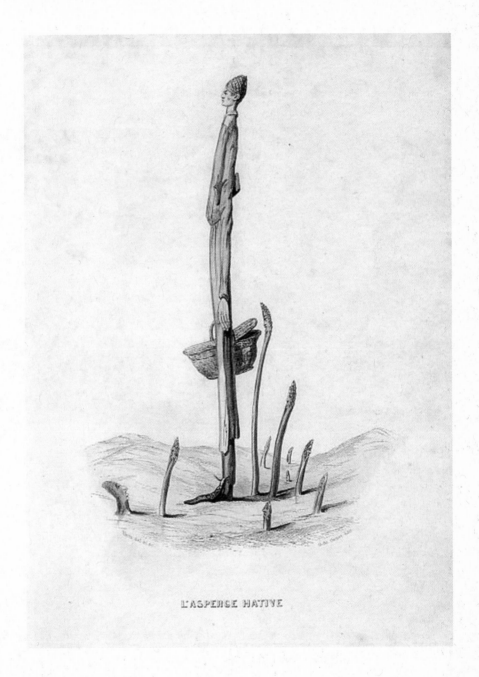

L'ASPERGE HATIVE

DOMINION REVISITED

RABBI ELLEN BERNSTEIN

R. de Salis

THE SIXTH DAY

And God said, "Let the earth bring forth a living soul after its kind: cattle and creeper and wild beast of the earth after its kind." And it was so. And God made the wild beast of the earth after its kind, and the cattle after its kind and every creeper of the earth after its kind; and God saw that it was good.

And God said, "Let us make adam in our image, after our likeness, and let them have dominion over the fish of the sea and the flyer of the heaven and the cattle and all the earth and every creeper that creeps on the earth." And God created the adam in God's own image; in the image of God, God created him, male and female, God created them.

And God blessed them and God said to them, "Be fruitful and multiply and replenish the earth and master it, and have dominion over the fish of the sea and the flyer of the heaven, and every live creature that creeps on the earth." And God said, "See, I have given you every grass-bearing seed that is upon the face of all the earth, and every tree that bears fruit, seeding seed; it shall be yours for food." And to every beast of the earth and to every flyer of the heaven and to every creeper on the earth that has a living soul, every green grass for food. And it was so.

And God saw everything that God had made and look! It was very good. And there was evening and there was morning, the sixth day. And God blessed them and God said to them, "Be fruitful and multiply and replenish the earth and master it, and have dominion over the fish of the sea and the flyer of the heaven, and every live creature that creeps on the earth."[1]

Va-yivarekh otam Elohim va-yomer lahem Elohim Peru u-revu u-milu et-ha- aretz ve-khiveshu-ha u-redu bi-degat ha-yam u-ve- of ha-shamayim u-ve-khol chayah ha-romeset al-ha- aretz.

THE HUMAN PLACE IN NATURE

In 1967, historian Lynn White Jr. argued in a now famous essay in Science magazine that the Bible gave humanity a mandate to exploit nature when it empowered the adam (human) to "master the earth," and "have dominion over" it.[2] Many environmentalists and theologians are still haggling over White's thesis, even after hundreds of articles and books have tackled the topic over the last thirty years.

In my environmental studies courses at the University of California, Berkeley, in the early 1970s, we read White's article and were taught that the theology of the Bible laid the ideological roots for the current environmental crisis. I naively accepted this idea, having no real knowledge of the Bible and no positive experience of religion. It was comforting to find a scapegoat to blame for society's problems, and religion has always been an easy target.

White's interpretation of Genesis had enormous ramifications on a whole generation of environmentalists and their students. I still encounter some who challenge my work, insisting that Judaism couldn't possibly have ecological integrity because "the Bible encourages people to control nature." They shun organized religion, claiming that it is the source of the environmental problem. It is conceivable that people who have little experience reading the Bible could examine this verse and decide that the language of "dominion" and "mastery over nature" is anti-ecological. But a verse is not a collection of words, just like nature is not a collection of plants and animals. Extracting a word or verse out of its context is like removing a tree from its habitat, taking it from the soil, the weather and all the creatures with which it lives in total interdependence. It would be impossible to really know the tree outside of its relationships. It's no different with the Bible. When you read the Bible, you have to consider the derivation of the words under consideration, the meaning of the neighboring words and verses, the message of the Bible as a whole, the context in which it was written, and how others have understood the verse throughout its three-thousand-year history.

The concept of "dominion" in this context is a blessing (*bracha*), a divine act of love. While God blessed the birds and fishes with fertility, God blessed humanity with both fertility and authority over nature. In more abstract terms the fish receive a blessing in a horizontal dimension while the adam is blessed in both horizontal and vertical dimensions. Like the animals, the adam is called to multiply and spread over the earth, but unlike the animals, he stands upright as God's deputy, overseeing all the animals and the plants.

Caring for Creation is an awesome responsibility. The psalmist captures the sense of undeserved honor that humanity holds:

177

What are human beings that You are mindful of
them
Mortals that You care for them?
You have made them a little lower than God,
And crowned them with glory and honor.
You have given them dominion over the works
of your hands,
You have put all things under their feet,
all sheep and oxen and also the beasts of the field
the birds of the air
and the fish of the sea, whatever passes along the
paths of the sea.[3]

As a blessing, responsibility for Creation is a gift. According to anthropologist William Hyde, the recipients of a gift become custodians of the gift.[4] The Creation is a sacred trust and dominion is the most profound privilege. It is necessary to remember the context of blessing as we examine the so-called "accused" words: *kvs*, "master," and *rdh*, "have dominion over." It is also important to remember that Hebrew is a more symbolic, multilayered, and vague language than English—any single word root can have multiple meanings, and often a word and its opposite will share the same word root. According to Bible scholar Norbert Samuelson, both *kvs* (master) and *rdh* (have dominion over) appear in these particular grammatical forms here and nowhere else in the Bible, so translating them is not a cut-and-dried affair.[5] The root of the Hebrew word for mastery, *kvs*, comes from the Aramaic for "to tread down" or "make a path." In the book of Zechariah, the root *kvs* is interchangeable with the root *akl*, the word for "eat." Although kvs is often translated as "subdue" or "master," it appears to have agricultural implications.

The root of the Hebrew word for "have dominion over," *rdh*, generally refers to the "rule over subjects." In a play on the word *rdh*, Rashi, the foremost medieval rabbinic commentator, explains that if we consciously embody God's image and rule with wisdom and compassion, we will rise above the animals and preside over, *rdh*, them, insuring a life of harmony on earth. However, if we are oblivious to our power and deny our responsibility to Creation, we will *yrd*, sink below the level of the animals and bring ruin to ourselves and the world.[6] If we twist the blessing to further our own ends, the blessing becomes a curse. The choice is ours.

As I was writing my book, I had long discussions with environmentalists and feminists who urged me to substitute a less "offensive" word for the word *dominion*, the traditional translation of *rdh*. They argued that *dominion* carries the negative connotations of control and domination. I considered what they said, and pondered the nuances of other words, like *govern* or *preside over*; one feminist suggested *have provenance over*. I decided that while these words are less offensive, they are also less inspired; they do not carry the sense of dignity and nobility captured by *dominion*; they do not capture the sense of taking responsibility for something much larger than oneself.

Like the Hebrew *rdh*, *dominion* implies two sides: graciousness and domination. Dominion, like money, is not in itself bad; it all depends on how we exercise it. As Rashi said, we can recognize our responsibility to nature and rise to the occasion to create an extraordinary world, or we can deny our responsibility and sink to our basest instincts—dominating nature—and destroy the world. Such is the human condition. It is time that we understand our conflicting tendencies and deal with them, rather than deny their existence.[6]

Humanity's role is to tend the garden, not to

possess it; to "guard it and keep it," not to exploit it; to pass it on as a sacred trust, as it was given. Even though we are given the authority to have dominion over the earth and its creatures, we are never allowed to own it, just like we can't own the waters or the air. "The land cannot be sold in perpetuity."[7] The land is the commons, and it belongs to everyone equally and jointly. In the biblical system, private property does not even exist because God owns the land and everything in it. When the state of Israel was established, the Jewish National Fund took responsibility for the management of the land—with an original intention to insure its perpetuity.

The blessing of mastery over the earth calls us to exercise compassion and wisdom in our relationship with nature so that the Creation will keep on creating for future generations. We use nature every day in everything we do; nature provides our food, shelter, clothing, energy, electricity, coal, gas. "Mastering" nature involves determining how much land and which animals should be designated for human use and the development of civilization, and what should remain untouched.

According to Sadia Gaon, in the eleventh century, "mastery" of nature meant harnessing the energy of water and wind and fire; cultivating the soil for food, using plants for medicines, fashioning utensils for eating and writing, and developing tools for agricultural work, carpentry, and weaving. It meant the beginning of art, science, agriculture, metallurgy, architecture, music, technology, animal husbandry, land use planning, and urban development.[8] That the power is in humanity's hands is clearly a risk for all of Creation. Indeed, the rabbis question why God created humanity, with the capacity to do evil, in the first place. Some of them figured that

humanity would only destroy itself and the world. But our ability to choose between good and bad is what makes us human. Free choice is what distinguishes us from animals, who follow their instinct, and angels, who have no will of their own and act entirely on God's decrees. It is up to us to determine if we will make of ourselves a blessing or a curse. To rule nature with wisdom and compassion is our greatest challenge, our growth edge. It demands that we understand ourselves and guard against our own excesses and extremes; it demands a constant level of heightened awareness.

One of the pleasures of grappling with a biblical text is that one can always find new meanings in it. Over the years, as I've turned this verse over and over, I've discovered a psycho-spiritual nuance. The complementary pair of blessings, "fertility" and "mastery," can be understood as blessings for "love" and "work." Fertility implies love, creativity, and being; mastery implies work, strength, and doing.

For most of us, love and work are the two dimensions that define our lives; for Freud they set the criteria for a healthy life. The complementary pair, love and work, take other forms, such as being and doing, sex and power. God blesses us with the ability to experience both. Yet our contemporary worldview attributes more value to our dominating side, to work, than to our fertile side, to love. It's important to temper our dominating tendencies with our fertile creative ones, and to remember that mastery over the earth is a sacred act, just like love is. They both invite the Divine in us.

The Buddha

Le Buddha

Je m'a mené sour les eaux. J'en savais plus que les docteur

gravure 'Flaubert'

NOTES

1. Genesis 1.28.
2. Lynn White Jr., "The Historical Roots of Our Ecologic Crisis," *Science* 155:3767 (March 10, 1967), 1203.
3. Psalm 8.4.
4. Lewis Hyde, *The Gift: Imagination and the Erotic Life of Property* (New York: Vintage Books, 2007).
5. Norbert Samuelson, *The First Seven Days* (Atlanta: Scholars Press, 1992).
6. Rashi, Pentateuch, and Rashi, Commentary, Genesis 1:26.
7. Leviticus 25.23.
8. Rav Saadia Gaon in commentary on Gen 1:26, quoted in Claus Westerman, *Genesis I–II, A Commentary*, trans. John Scullion (Minneapolis: Fortress Press, 1984).

From Ellen Bernstein, *The Splendor of Creation: A Biblical Ecology* (Cleveland: Pilgrim Press, 2005). Used with permission of the author.

THE FIRST RED TOMATO

BERNADETTE DIPIETRO

Mr. and Mrs. Valentino had the best corner in the neighborhood for evening summer games. It was ideal for "hide and go seek" and "tag on sewer lids." The corner had one sewer lid on each curb and three additional lids on the street, but most importantly, it had a streetlight. Frantically we would run from sewer lid to sewer lid, hoping that the person who was it would not tag us until we stepped onto the next safe lid. We yelled as we ran from lid to lid, we screamed, we laughed, and we played so hard, until our mothers called out of the darkness to come home.

Mr. and Mrs. Valentino lived on that corner in the red brick house with green and white striped awnings extending well beyond the walls of the red brick porch. Their house was newer and sat higher than the row of brownstones Across the street. When Mr. Valentino and his wife, Millie, were sitting on the porch, one could only see the tops of their heads from the street below. There was an unspoken sense that they were quietly watching us as we played our games into the late summer evenings. My grandfather was a friend of the Valentino's and spent many evenings playing cards over glasses of wine as they shared the Italian language that was unfamiliar to the other nationalities in our neighborhood. I always imagined as a child that they were talking about their two loves, gardening and winemaking.

Next to our apartment building were two of the most magnificent gardens in the neighborhood, tended by my grandfather, Giuseppe. These gardens sustained our large extended family for the entire summer. He grew beautiful red and green chard, eggplants, red and green peppers, red leaf lettuce, green beans, basil, squash, zucchini, and delicious red tomatoes. His garden was planted by the moon. Each year, grandfather would receive for Christmas a large Mellon Bank calendar with tiny images showing the phases of the moon. He would post it on the back of his bedroom door and plan the next season. He would start his plants from seeds in empty milk cartons on his bedroom windowsill when the snow was just beginning to melt, in preparation for spring. His garden soil was rich, soft, and black like none that I have ever seen in any part of the world. When spring finally arrived, grandfather would devotedly carry those tender young plants outside in his large Italian hands and marry them into the rich, black earth.

The two gardens were located on either side of our apartment building. How wonderful to spend time with grandfather, singing as I picked weeds and he planted. He was a tall man with a beautiful white mustache and a tender loving smile. He had a strong presence, was clever, had a fine sense of humor, and was always a step ahead of everyone else. He wore white starched shirts and black ankle-high laced shoes that only the European men wore in our neighborhood.

Mr. Valentino, on the other hand, was the opposite of my grandfather. He was shorter and stockier, and wore white starched shirts and big-bellied pants with large black suspenders. Mr. Valentino and my grandfather were antagonistic friends. Each spring they would try to outdo each other with their gardens. It was the springtime, summertime competition. Let the games begin!

Unlike Valentino's gardens, our gardens were visible to the neighborhood. They were visible from both Second Avenue and Chatsworth Street. So Mr. Valentino had easy visible access to the daily growth of grandfather's plants. Each morning he would religiously pass by the garden in his white pressed shirt, pipe dangling from his mouth, and peer over the fence. It was the

daily visual measurement with comparison in an Italian format.

Grandfather despised this intrusion. He would comment during our evening family meals about Mr. Valentino passing by the garden with his relentless eyes. After many summers of such meddling, grandfather had enough and decided to take revenge with a humorous approach and a solution. He asked me one evening to borrow one of my small red bouncy rubber balls that I used for playing jacks. At dusk he set out to tie the small rubber ball ever so gently inside one of his tomato plants. He placed it deep enough into the bushy plant so that it would be particularly visible to the hawk eyes of Mr. Valentino, standing at the fence between Chatsworth Street and grandfather's garden.

Early the next morning grandfather sat by the window with pleasure and delight as Mr. Valentino surveyed the first ripe red tomato of the season in my grandfather Giuseppe's garden.

Nick Hayes

BAKING

GRAISON S. GILL

for Andrew, Blair, and mostly Phineas

Baking is a humble endeavor within a fabric of increasing exaltation. We live in a city of personality—character of characters, petty gentry; it's as if God hurled a lopsided kaleidoscope blind to earth. The refraction of its crashing light impregnated every mosaic tile that built New Orleans. Broken tesserae that, individually, lack coherence. It's through gentle lapidary and mutual assemblage that they become cohesive. The image becomes phonetic, the writing on the wall legible; like the arrival of vowels at a party of consonants. And, like those tiles, not even the dead, or our tree's roots, are kept underground in this city. So that we may not lose sight of them, and afterward ourselves. The fear of concealment precipitates blindness; as Rilke said, God walks through men in the South.

Our new mill, in its own way, provides the mosaic an adhesion: in a place where, unfortunately, culture created by the exterior—the underclass, subaltern, marginalized, oppressed—is being liquidated into "capital" and materialized into a "cultural economy." (Consent has been manufactured, and culture is hiring a foreman). We believe sincerely in the integrity of our project. We never seek to reduce, revise, renege, develop, invigorate: nothing about New Orleans needs a shot in the arm. Only a fool tries to change what he loves; and his behavior will only change him. If you're trying to change New Orleans, go home. So, with all the hypocrisy and pandering of Katrina 10 over, we can focus on life, here, now. Anniversaries, like funerals, are postcards from the present. Beveled mirrors in simple frames: don't look into mirrors to see in front of or behind yourself. "Joy and peace are the joy and peace possible in this very hour.... If you cannot find it [here, now], you won't find it anywhere."[1]

I find purchase in bread. Always have, since day one. And there is immeasurable, inconceivable value to milling our own flour: nutrition, flavor, freshness. And we truly desire to remind anyone who will listen that we used to have this proximity in every walk of life. The baker to his flour, the musician to her sax, the cork to its bottle, New Orleans to its perpetual past. But that elegance is gone, that grace, its poise— the ear now bankrupt of the word beauty in all its robes. Values are not ephemeral; the Revolution of the sixties failed because people tried changing others before changing themselves.

By doing our part, in fulfilling selfishly our calling, we hope that our clarity will empower the bigger picture. An act of love as simple as milling wheat into flour. Think of it. It's banality is superb, deceiving; recorded history's first ink dried here, with one person sitting down to mill grain in Baghdad. And look where we are today, still, well, making music, milling grain, divorcing, talking about ourselves. Nothing has changed the space in between, mere parentheses.

NOTES

1.   Thich Nhat Hanh, *Miracle of Mindfulness* (Boston: Beacon, 2016), 36.

GRACE

JASON BENTON

He saw a Samaritan who was trying to take away a lamb while he was on his way to Judea.
He said to his disciples, "That man is pursuing the lamb."
They said unto him, "So that he may kill it and eat it."
He said unto them, "As long as it is alive he will not eat it, but only when he has killed it and it has become a corpse."
They said, "He cannot do it any other way."
He said unto them, "You, too, look for your resting place so that you may not become a corpse and get eaten."
—Jesus and his Disciples, The Gospel According to Thomas

v

Saying grace has been such a habit of eating that it never required thought. Before you eat, you say Grace. I'm far from the regular churchgoer I once was, but I still develop a bit of a hiccup if I start eating at a table without the opening ritual of prayer and thanks. However, since I started raising, slaughtering, and butchering my own meat I have started to think about it. Grace has become grace. Grace has become place.

RETURNING HOME

We have a small property—not big enough for cows, but just big enough for a small herd of goats, a couple of pigs, a flock of hens, and a large garden. My wife and I returned to our rural roots after more than a decade in the city: college, jobs, grad school, and more jobs. When we had our first daughter we decided home and family was the best place for our children. And I've always had fantasies of farming, homesteading, and living off the land like my great-grandparents and ancestors had. But like most people my age, we are separated by a generation or two from the land, from that life, and for most of us it is long forgotten to lives devoted to career, city, and debt. Yet I still dream about it.

So once I had my small plot of land I started asking my grandparents about their wisdom and knowledge of farming. And of course, as trained by my decades of education, I read books on goats. Some very good books are written of goats and goat herding, and from these I've gotten the idea that people do not breed goats—goats breed a certain character of human being. One such book, written by Brad Kessler, made me nearly quit my job and move to Vermont. But reality keeps me planted and employed where I am for now—the same reality that keeps me from being vegetarian or vegan. Meat is convenient but also fun, delicious, and nutritious. It is sun, atmosphere, and earth stored in flesh. Because I feel this way I decided to make responsible use of my pasture of grass, scrappy trees, and brush. Instead of mowing it (don't get me started on the American lawnmower obsession), I'll turn it into meat—my meat—not the stuff from who-knows-where sitting on the soggy absorbent blooming pad at the grocer.

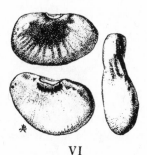

VI

DREAMS

Dean, my great-grandfather, and Clifford, Grandpap, were tearing out multiflora rosebushes and thinking, dreaming, scheming half aloud while they grunted and sweat. Dean had drawn up some plans and had a local blacksmith build a contraption that could be pulled by truck, man, horse, or tractor on one end, and an opposing lever would engage a pick that would thrust under the roots and grab hold of the bush, pulling it out of the ground almost effortlessly, root bundle and all. They were putting it to the test that day, ripping out thorn bushes with gusto. But it was still too much work for a worthless bush. Some of those bushes had such deep and wide root systems that they left a divot the size of a pig, scarring good pasture for their dairy farm. One of them had the idea: goats. They eat the poison ivy, thorns, multiflora, and all the other brush that they have toiled to keep out of the pastures. Goats are friendlier than pigs, less expensive than cows, and reproduce reliably. They would branch out, create a market for chevon and goat's milk, and keep working their land. Getting rich wasn't part of the plan—just to continue to exist and enjoy the place they called home. It was a perfect plan, as all plans seem at their genesis.

But then they were "kicked out," as they say. The government, mines, and corporations moved in, the land disappeared, and there was no more farming. Cliff joined the Army and left. Dean worked in the coal mines until he died of black lung. After 175 years or so of farming, the surface of the land was strip-mined, then let loose to its relentless path back to Pennsylvania forest—as a lovely state park where I now play with my children.

These plans were being dreamed about twenty-five years before I was born. I learned of them when I told Grandpap, Cliff, about my own dreams and schemes of raising goats for my personal sanity and consumption. He lit up, immediately giving me tips, talking about breeds, fencing, milking, breeding, and such. He quickly calculated how many goats my land would sustain, how much hay I might need for the winter, how to compost the manure for gardens, what the meat and milk should taste like, and how much milk I should produce in a week. I can't recall the last time I had such a stimulating conversation, the last time I freely dreamed and schemed and had purpose—good purpose.

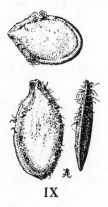

IX

E-I-E-I-O

Why does farming have this effect on us, or at least some of us? Why does small scale, light, sustainable farming have such allure and become the conclusion to the ailments of our civilized ways? Kids who will never set foot on a farm play with farm toys, and we tell them agrarian rhymes, fairy tales, and lullabies. As soon as somebody around here strikes it rich with gas royalties, they go out and buy some more land and a tractor. This is such a trend that John Deere opened a new dealership nearby. But of course, that bubble has already burst.

In our town there is a romantic mural depicting the agrarian heyday of this area, and we all seem to celebrate it through our antique fetishes, urbanites visiting quaint farms, and the upper-middle-class folks weekending in places where the locals do not have the time or money to vacation because there's a harvest to get and mouths to feed. Spiritual gurus like Steiner and many ecologists have come to the conclusion that sustainable small-scale farming is essential to life if we are to continue living good lives. Writers like Marvin Bram, W. Berry, and many others have also placed farming, gardening, and husbanding on a high pedestal as a remedy to civilization's mistakes and ills.

I'm a descendant of farmers on both sides, so maybe my blood creates my bias. The back-to-landers made their attempts at an exodus during the couple of decades prior to my generation's birth. While many of us still share their sentiment, we can see how, in many ways, they failed—out of romanticism, naivety, and impracticality. Escape is a nice idea and a constant temptation, but it's impossible and maybe irresponsible. Perhaps without a keen business sense, without working with the Man in at least some ways, without using at least some of the perks of civilization (i.e. money), farming at any scale is not sustainable unless you have a community and local economy to support it. And good land is harder and harder to come by, especially if you're not already money or land rich. Furthermore, to live on the land, with the land, by the land, one must think in terms of soil, water, and atmosphere in the span of centuries and millennia, but always within the locale. This moment will soon be ancient history or maybe forever forgotten.

This land must be better once we leave it—that is our primary investment.

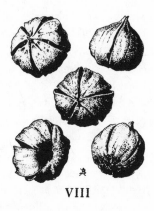

VIII

THE MACRO VERSUS THE MICRO

That is the paradoxical outcome of our civilization: that which is necessary for life and methods for sustainably supporting life—all life, including wildlife—are practically unsustainable because of the deeply ingrained and rigid structures of economy, trade, labor, and land management. Small-scale farming does not have any profits as an end goal, nor should it be co-opted as such, as industrial commercial agriculture has. Small-scale farming has nothing to do with the global condensation of capitalism—feverishly gleaning natural resources, including human minds and lives, to produce capital, namely money, and uncontrolled growth of wealth and human population. I'm reminded of Edward Abbey's "Growth for the sake of growth is the ideology of the cancer cell."

Such unrelenting and unfettered desire will ultimately ruin the object of desire, the greedy one, and all in between. This has been clearly evident for some time now in our environment, the extinction of species, climate change, epidemiology, and the human spirit. Everything is turned into an industrial factory: education, churches, and health care produce human capital; forests, fields, and oceans provide the food, mining, and fiber. The information tech revolution is just another piece of technology that aides and abets this so-called democratic capitalist machine—in many ways it may be yet another extension of large-scale industrial agriculture. This has begun to harvest the minds and consciousness of humanity, creating one giant network of consuming robots (corpses?) constituting one giant consuming robot. Perhaps this is why we have been so entranced by zombie stories.

IV

RESURRECTING KNOWLEDGE

Nevertheless, I'm astonished and somewhat puzzled by my grandparents' adaptability and Zen-like acceptance of the way things are, even if they are fundamentally opposed to it or terribly burdened by the changes in the world of their lifetime. Resentment never soured their spirit or generosity. They talk of the past with reverie but without longing or bitterness. They've always been grounded in the present and the work at hand. They have kept about their business of caring for their family, neighbors, and property with skill and grace, and this continues to sustain them.

My education had been quite different. It was laden with a selfish anxiety over all I was supposed to do, and all the while with the knife of poverty at my throat: get good grades, graduate college, get a good job, make gobs of money, then retired and die. This was very much part and parcel of my public education, which did not include the Grange or FFA as my grandparents' schooling had. There was an inherent, subversive message just under the surface that I would not be worth anything if I did not attain these things, and this message drove me and many of my classmates to the cities for college and careers, where most of us have remained.

My grandparents' teachings were imbued with an urgent responsibility to preserve and care for their land, family, and community—much of which is now gone or dispersed in one way or another. My postmodern message, luckily, wasn't pressed by my parents, although I felt their anxiety over my future. Like many rural families, they struggled for a while with unemployment, fearing homelessness and the shame of being "on food stamps" and the recipient of local charity. I now know a wealthy CEO with the same anxiety: without constant growth, innovation, and improvement, then utter failure and depravity. It is all or nothing today—bipolar.

But for my grandparents and their parents, the goal wasn't riches or bust, but simply to get by, and just getting by wasn't that bad. It was possible to be poor, to live without money, without being depraved, shamed, and starved. It was possible to get by with little but still have a meaningful and rewarding life. A typical response would be, "Well, they had to work very hard for what they had." That was true, but look at how much

people work today, even to the point of injury, illness, death, or suicide. Free time and leisure are rare, and many rural households work two to four jobs and are still under a great deal of debt and stress, just a paycheck away from foreclosure or bankruptcy. For multiple generations now, rural parents, the descendants of farmers, miners, and millworkers, in that respective chronological order, are forced to spend most days away from children to pay the bills and the bank. Addiction, namely heroin, is a rural epidemic, and overdoses are now the leading cause of accidental death in our area. Hobbies are becoming an activity of the upper-middle and upper classes only. Most farmers I know have one or two people on the farm working in mills, mines, or other places in order to maintain the farm. Many find it easier to subdivide the farm, sell or lease to miners or drillers, and most are doing just that. I recently heard a Pennsylvania politician say something to the effect that in order for family farms to survive, they need to have a gas or oil well on their land.

So what does it mean when a farm needs to produce less food and more fossil fuels in order to exist? We know who benefits and who doesn't from this plan. What should a "farm" be? What should a rural home be, in relation to local and global commerce, to the city, to other rural homes? And as Wendell Berry asked, What are all of these people for?

FAMILY

Her face was in her hands, tears leaking through her fingers and dripping onto her plate of lemon chicken and garden sugar snap peas. Pig-tailed and rosy-cheeked, her large brown eyes peered over her hands, "But they won't be for meat, Daddy ... will they?"

I had been reading about goats: breeding, raising, milking, cheesing, browsing, haying, fencing, butchering, freezing, canning, jerkying. My five-year-old daughter was looking over my shoulder one day while I read. On the page were a series of photographs demonstrating the sticking, exsanguination, evisceration, hanging, and skinning of a goat. She was very quiet, and after putting it all together said very plainly and firmly, "We won't be doing that to our goats. We will just milk them and play with them."

Over the winter months we had ongoing conversations, usually over the dinner table and whenever we ate meat, about our plans for goats. Of course yellow and purple carrots are fun, and frozen or canned corn, beans, and tomatoes are a garden staple, but she especially liked lemon or curried chicken and anything with bacon was "really good." As spring approached and the kids were being born on a nearby farm, our five-year-old daughter became more excited about the prospect of frolicking baby goats in our brushy backyard and flowery field. She drew and painted pictures of goats, thought of names (Anna, Dalia, Lahla, Gloria), goat games, goat clothes, goat parties. She couldn't wait.

"But their mommies will come and visit because they'll miss them, won't they?"

"Will they miss their mommies? Will their mommies miss them?"

"Do they know they're going to die?"

"Do they know they're meat?"

"Do they know we're going to eat them?"

It was a lot for her to take in. As for our three-year-old, it was all very simple: "I don't eat animals, I eat meat. Duh." And for her three-year-old cousin, he had to let his father and me know about our obvious blunder while we stood over a deer his father had just killed. We were discussing our plans for the dead animal, including steaks, sausage, roasts, and jerky. "We don't eat animals, silly daddy!"

I

FOOD

Well, we never like thinking about it that way—at least I don't. I enjoy animals. I love animals. There is little I enjoy more than watching a wild animal in the woods or relaxing with a domesticated creature: a doe eating fiddleheads, a duck tending to her ducklings, a groundhog scanning the horizon, a rabbit grazing, a rooster marching across the yard, a hen perched on my shoulder, a cow licking my hand, a pig basking in the sun or grunting with pleasure after I scratch her ear, a frolicking goat. To commune with, care for, and play with animals makes life more meaningful, more worth living. After all, I'm an animal. But I'll admit, killing and eating animal is haunting, hard work, and indeed delicious.

The killing is never enjoyable. Calm, quick, and efficient. Hoof to hanging meat as fast as possible. Once the animal is a carcass, it's easy. (Note to reader: if interested, be sure to cut the horns off the dead goat before skinning—it will make skinning and beheading at the atlas much easier). Use the strength in your fingers. Let the knife follow the bone, no waste. Liver and onions. Chilled and thinly sliced cardiac muscle. Smoked bacon. Sweet breads. The mason jars of roasted and boiled bone broth lined up on the basement shelf. A full freezer. My first slaughter and processing of a goat went very well thanks to a butchering manual by Adam Danforth, which I highly recommend.

But the killing and cutting also reminds me of my own mortality, my own meatness. I actually feel I learned much more of my own human anatomy and physiology than I learned in an entire year of the graduate course. I am reminded that I am animal—with breath. I am food too. Dust to dust, food to food, fodder to fodder. I am life—and life is abundant yet sacred. I am responsible for the lives of others and of the earth.

We named our first meat goat Jerky to remind us of his fate. Once in the freezer, the girls asked to see him. They seemed a little disappointed not to see horns, hooves, and fur leap out of the freezer door—a reasonable expectation given that goats, especially Jerky, find their way on top of or inside just about anything.

So when I purposefully kill an animal, and I've killed many, including fish and pheasants, grouse and geese, ducks and deer, fish and frog, there is both a sense that this is something I'm good at and something I resent. Railroad Earth has a wonderful song, "The Hunting Song," which captures the emotion well: "I felt it way down inside." What do I resent? Death? Senescence? Or the violence of life? This resentment does not foul the flavor of meat for me, but does encourage me to eat much less meat and savor the meat I choose to eat. It nudges me toward a contemplation of this place and a thoughtful and loving consideration of my plants, animals, and myself. Killing and processing every little morsel is also hard yet gratifying work. An entire year or two of attention, care, sweat, heartbreak, and annoyance—my wife and I in the barnyard in the middle of a rainy night restraining and administering a homemade plant-poison remedy to deathly sick goats spraying us with volumes of projectile vomit—condensed down to sixty or so pounds of edible meat, bone, and organs.

Work, play, food, home, and relationship—even life and death—become one.

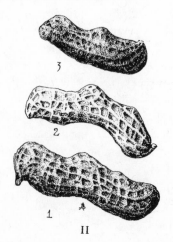

192

II

LAND

There will be frolicking bleating kids this spring, making happy children.
Milk for yogurt and cheese in the summer, making good gifts.
Meat in the fall making hearty meals year-round.
And more kids in the spring.
And more milk.
And more meat.
And more kids ...
My children will grow
To know
This land.

III

Thank you for all the land and life this place has to offer, all the wonderful life forms, energy, death, and hard work put into our food. Bless it all to our bodies and spirits ...

And thank you, goat.

Amen.

Martha Shaw

JULY

LABOR AND AIRWAVES

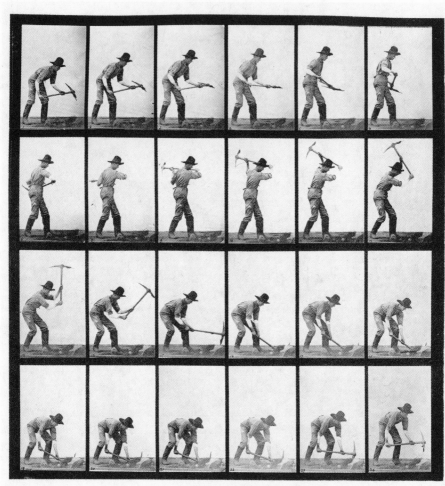

Eadweard Muybridge

I sometimes fear the younger generation will be deprived of the pleasures of hoeing.

—John Updike

WILD SEEDS

MARÍA JOSÉ GIMÉNEZ

She grew wild tomatoes—the original small fruit that frightened the conquerors. She harvested and ate them a handful at a time as they ripened. They never made it to the kitchen except as seeds squeezed out to dry on a paper towel for anyone who wanted to plant them.

She foraged bitter brown pods whose name is lost in the recesses of childhood in another tongue. Ready to be chewed for anything that ailed the gut, they sat on the windowsill just outside her walk-in closet of a kitchen, a dark space with a tiny window and just enough room for one person to stand in front of the four-burner gas stove.

She cooked early in the morning, before the stifling midday heat, left giant pots of soup or stewed meat from the ranch in case people showed up for lunch.

And they did. Summers were peppered with surprising numbers of new cousins. They would walk in through the long zaguán after knocking on the metal door with a coin or a key. And she would introduce them: "This is Fulanito, son of Fulana," and she'd name a woman whose name none of the cultivated cousins ever recognized or would remember.

She never mentioned who their father was or how they came to be relations. Time revealed more than a few uncles had had a fruitful extracurricular life and had sown the town with their descendants while on family holidays from the big city.

My grandmother fed and watered them all.

7. *Cardiaca.*   Motherwort.

$15 MINIMUM WAGE:

Disaster or Opportunity for Family-Scale Farms?

ELIZABETH HENDERSON

The growing momentum of the campaign to raise the minimum wage presents those of us who are farming with a serious challenge. How are we going to respond?

If the minimum wage had kept up with inflation, the $1.60 of 1968 would be $10.96 today, so workers' demands for raises are getting serious consideration. The fast food workers' Fight for $15 has pushed the New York Labor Board to back a plan to phase in this new minimum over the next six years.

To many organic farmers, $15 per hour looks good—as a wage for us farmers. As employers, however, it puts a lot of strain on our businesses. The question farmers need to ask is, How do we turn this into an opportunity? Can we inspire a campaign to raise prices for farm products so that we can act in solidarity with other food workers instead of joining the Farm Bureau litany: "This will wreck farming in New York State"?

If we want to make real progress toward a more resilient and sustainable food system, we have to do a much better job of linking justice for farm workers and justice for farmers. It is difficult to make the finances work on a family-scale farm. The number of farms in the United States has shrunk by over four million during my lifetime. When we talk about food justice, we are not just talking about something we need to do for others, for exploited farm workers or undocumented dish washers. Farmers need food justice too.

Would-be farmers need to understand that they will probably spend years as employees—working for other farmers, working as hired farmers. We need a system that supports fair pricing. Remember parity? Price supports? A minimum wage for farms? That ceased to exist in the 1950s. Farmers should know this history and join in the movement for justice as our own agents. Once you start to think this way, it is not too hard to see that farmers alone are not going to get very far. By cooperating and working in solidarity with other food system workers—17 percent of the entire work force—there might be a chance to get somewhere.

Most organic farmers I know are painfully aware that the current cheap food system coupled with "free" trade makes it difficult to keep family-scale farms afloat. Over the years since World War II, family-scale farms have been going out of business at a steady and alarming pace until very recently. In 1943, the year I was born, there were close to 7 million farms. There

organic processors, are poor. Most small farms are not profitable, and many are in debt. Legal protections that would allow farmers to form associations to negotiate contracts with buyers are weak.

The sustainable agriculture and local foods movements have reversed the downward trend in farm numbers, and the number of very

Silver ducat from the time of the last Doge of Venice

are only 2.2 million today. A major squeeze or speed-up has been underway in farming that has been especially hard on dairy farms and farms that produce commodity crops. Rising costs, the droughts and floods of global warming, and low prices due to concentration in markets that reduces the number of possible buyers have all contributed to tight budgets for farms. Prices of commodity crops, especially soybeans, corn, and wheat, have been under pressure because of expansion of large-scale export-oriented farming in countries such as Brazil, Argentina, Russia, and Ukraine. Contracts, even those made by

small farms is actually growing. Nevertheless, something like 84 percent of existing farms are in debt. Prices do not cover farmers' costs of production. Many of the farms that do not hire labor do have a family member who works off the farm so that the farmer can have health insurance, send children to college, and save for retirement, or the farmer works a regular job and spends evenings and weekends doing farm work. While there are some outstanding examples of farms without much hired labor that are doing well financially, most of the family-scale farms I know are struggling to make ends meet, or are run by people who have chosen to live "simply." Often, farmers are so discouraged about the money aspects of their farms that they do not even try to calculate costs accurately. They farm for the love of it, and either eke out a living that would qualify as below the poverty line or make money doing something else to support their farming habit. Family-scale farmers are fragile small businesses, a marginal population in the United States and all of North America.

As farm commodity prices fall again, more farms are likely to go under. At the end of August 2015, the U.S. Department of Agriculture (USDA) predicted that farm incomes will drop to less than half the peak reached two years ago. The USDA projected farm incomes this year will come in at less than $59 billion, down 36 percent from last year and 53 percent from a record high of $123.7 billion two years ago.

By contrast, the organic market is growing quickly, though that does not translate into profitable sales for U.S. organic farms. According to the USDA Census of Organic Farms, "63 percent of U.S. organic farms reported selling products to wholesale markets. These sales accounted for 78 percent of U.S. organic farm sales. Wholesale markets, such as buyers for supermarkets, processors, distributors, packers, and cooperatives, were serving as the marketing channel of choice for U.S. organic farmers to get organic agriculture products to customers." The census shows that sales of organic crops and livestock at the farm gate reached $5.5 billion in 2014, up 72 percent from 2008, and according to the Organic Trade Association, overall organic sales in 2014 were $39.1 billion, up more than 11 percent from the previous year. In analyzing the census results, Edward Maltby of the Northeast Organic Dairy Producers Association points out that the number of organic farms and acreage have been shrinking: "There were 14,093 organic farms in the United States last year, accounting for 3.6 million acres, with another 122,175 acres in the process of becoming organic, according to the latest National Agriculture Statistics Survey. However, in 2008 there were 14,540 organic farms making up 4 million acres with another 128,476 acres going through transition." According to Maltby, there has been attrition even in organic dairy farms as they get caught in the squeeze as input costs rise faster than organic milk prices.

Processors are importing more organic ingredients. In part, this is to make up for the insufficient expansion in U.S. organic production. The lower prices of imported organic crops, however, may be more important than the U.S. shortage, and the availability of cheaper imports creates a downward spiral, discouraging U.S. conventional farmers from transitioning to organic. Klaas and Mary Howell Martens identify exchange rates as an additional factor: "A big factor in organic imports right now is the currency exchange rates. One Canadian dollar equals about $0.75 U.S., making a $20-per-bushel grain in Canada convert to $15 U.S. The sharply rising value of the American dollar causes commodities to fall equally sharply. That makes importing grains very profitable for brokers. Anybody living near the Canadian border right now can find lots of cheap grain being offered to them from up north. Eastern European and South American grains are equally cheap to import right now."

Three years years ago, the Domestic Fair

Trade subcommittee of the Northeast Organic Farming Association (NOFA) Interstate Council did a survey of organic farms to learn what labor policies they use and what wages and benefits they are paying hired workers.[1] The survey disclosed that most labor is done by the farmers themselves and their families. Those who hire labor expressed the desire to pay good wages and benefits, but most were paying no more than $9 or $10 an hour. Many commented about the obstacles to living their values—low pricing, lack of markets.

Three members of the board of NOFA-NY have been involved in a dialogue with Rural Migrant Ministry and other members of the Justice for Farmworkers Campaign about the Farmworker Fair Labor Practices Bill, the focus of a multiyear campaign. NOFA-NY polled NOFA member farmers to find out which provisions of that bill are acceptable and which ones farmers would like to see modified so that NOFA members can wholeheartedly endorse the bill. NOFA-NY plans to continue this dialogue and broaden it to include more stakeholders.

I would hope that we organic farmers can agree on a long-term vision that will solve our economic stresses together with the problem of who will be available to work on our farms in a holistic and humanitarian way that honors the organic principle of fairness. At stake here is what in policy discussions of labor supply is being called the "future flow"—who will be allowed to enter the United States and under what conditions. For the past half century at least, industrial-scale agriculture has depended on a steady supply of immigrants from other countries. Nowadays some of the small farms

that employ labor also depend on immigrants. Most of those who have come to work on the farms in the United States are people who had been driven from their own land by economic hardships and political upheavals.[2] The free trade agreements, especially NAFTA, have increased the numbers of desperate farming people coming to the United States from Mexico and Central America. Most of them would not come here if they could make a living on their own farms back home.

Do we want to imitate the labor practices of monocrop industrial agriculture? The structure of those farms resembles, in a chilling way, the structure of the old slave plantation. Is this what we want for the organic and agroecological farms of the future? I suggest that instead, we highlight that smaller-scale highly diversified farms are more productive per acre and also offer higher quality, more diverse work. On a farm like mine, work is varied and interesting; you rarely spend eight hours in a row doing the same task. Besides, organic methods—rotations, cover crops, soil-building through compost, reduced tillage, biodiversity—buffer the effects of climate change, with less exposure to toxics for farmers and farm workers, and for the soil, the environment, and customers.

For our farming to be worth sustaining, farm work must become a dignified, respected career path, properly remunerated with a good benefits package. The farm becomes a center not only of production but of training and cooperation with the community.

With our eye on the long term, we must shift from an agriculture that depends on the constant influx of desperate low-paid workers to a domestic workforce. We need to elevate farm work to the place of respect it deserves. That means pushing in every way we can imagine to remunerate farmworkers adequately, starting with the farmers ourselves. Everyone working on our farms deserves living wages with decent

cooperative structure, whether that is several members of a family, a formal legal cooperative, or a group of farmers working together.

Whatever merits capitalism may have in other sectors, the laws of corporate capitalism are totally inappropriate for food production,

Fifteenth-century Labors of the Month

benefits, health care, retirement, and funds for professional development. When new immigrants arrive, they should not be regarded as a source of continuing cheap unskilled labor but, if they want to stay in the United States, as additional recruits and reinforcements for our campaign for new farmers.

We need to provide access to the resources the new generation of would-be farmers requires so that they will be successful. That means higher farm-gate prices, and access to land and credit. There needs to be a diversity of farm-related jobs—not everyone wants to be a farm owner-manager. Farms operate better with a more

distribution, and sale. The food system of a new economics will have to be fair, apportioning the food dollar up and down the food chain— or throughout the food web. What we need is domestic fair trade, where buyers pay farmers enough to allow them to use sustainable farming practices, to earn a living wage for themselves and their families, and to pay living wages for the people who work on their farms. Living wages include shelter, high-quality culturally appropriate food, healthcare, education, transportation, savings, retirement, self-improvement, and recreation. The Agricultural Justice Project has assembled farmers, farm workers, and other stakeholders to compose high-bar standards for fair pricing, and decent working conditions for people who work throughout the food system that is a useful guide for fair trade.

To fulfill the vision of feeding our population locally, we need many more farms and many more farm workers. This will require radical redistribution of land—either breaking up big holdings or creating land trusts by county or state, holding the land in common and leasing it to people who work it—a return to usufruct. And the public will need to change their diet to rely to a greater extent on what we can grow in our region—a new mixture of annuals and perennials.

So how do we get there? If we want our movement to have the strength to replace the industrial food system, we farmers need to work as allies with all the other food workers, from seed to table. Despite owning significant amounts of land and equipment, the earnings of farmers like me and many of you are more like those of industrial workers than captains of industry. The profits in the food system go almost exclusively to the other sectors: "the agricultural family unit is only a subcontractor caught in the vise between upstream agro-industry … and finance … and downstream … the traders, processors, and commercial supermarkets."[3] Family-scale "sustainable" farmers will only break this vise by taking our place alongside other working people in the food system in solidarity with their struggles that are really our struggles as well.

If we at least take our stand with other food workers and begin demanding that farmers, farm workers, and all food workers make living wages with full benefits from a forty-hour week, we may start moving toward an agriculture that will sustain us into a future worth living. Local organic agriculture should serve as a model value chain, changing relationships to bring alive the Principle of Fairness that is fundamental to organic agriculture all over the world. Let's take the opportunity that public attention to raising the minimum wage presents to us and raise our voices for justice for farmers and farmworkers together.

THE PRINCIPLE OF FAIRNESS

Organic agriculture should build on relationships that ensure fairness with regard to the common environment and life opportunities.

Fairness is characterized by equity, respect, justice, and stewardship of the shared world, both among people and in their relations to other living beings.

This principle emphasizes that those involved in Organic agriculture should conduct human relationships in a manner that ensures fairness at all levels and to all parties—farmers, workers, processors, distributors, traders, and the general public. Organic agriculture should provide everyone involved with a good quality of life and contribute to food sovereignty and reduction of poverty. It aims to produce a sufficient supply of good quality food and other products.

This principle insists that animals should be provided with the conditions and opportunities of life that accord with their physiology, natural behavior, and well-being.

Natural and environmental resources that are used for production and consumption

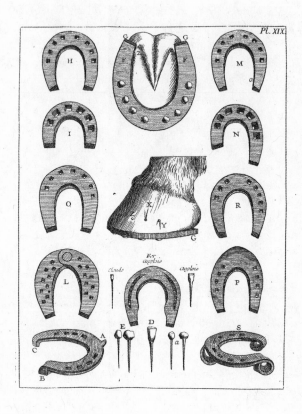

should be managed in a way that is socially and ecologically just and should be held in trust for future generations. Fairness requires systems of production, distribution, and trade that are open and equitable and account for real environmental and social costs.

NOTES

1.   Becca Berkey, "Just Farming: An Environmental Justice Perspective on the Capacity of Grassroots Organizations to Support the Rights of Organic Farmers and Laborers," January 2015, http://www.northeastern.edu/nejrc/.

2.   See the film *Harvest of Empire: The Untold Story of Latinos in America*, directed by Peter Getzels and Eduardo López, 2012.

3.   Eric Holt-Gimenez, *Food Movements Unite!: Strategies to Transform Our Food Systems* (Oakland, CA: Food First Books, 2011), xii.

Originally published as Elizabeth Henderson, "$15 Minimum Wage: Disaster or Opportunity for Family-Scale Farms?," *Food First*, December 19, 2015, https://foodfirst.org/15hour-minimum-wage-disaster-or-opportunity-for-family-scale-farms/.

RETHINKING THE VALUE OF WORK

JOHN IKERD

"How can it be that more than a century after muckrakers exposed the deplorable conditions of workers in the food system, that harassment of workers, rapes in the fields, squalid living conditions, pesticide showers, hazardous working conditions, and slave wages continue be the norm?"[1] In reviewing the documentary film *Food Chain*, Jane Kolodinsky provides this fitting description of the inevitable consequences of the commodification of labor in an unrestrained market economy.

The deplorable working conditions in the food industry have not been corrected because such conditions are inherent in the industrial system of food production. More effective labor unions and ethical choices by consumers might relieve some of the suffering—at least temporarily. However, the well-being of workers in the food industry and elsewhere will not be significantly improved until we rethink the value of work and restrain our economic system accordingly.

The most basic function of a free-market economy is to allocate land, labor, and capital among alternative uses so as to maximize consumer utility or satisfaction. Anything that needlessly increases the cost of food to consumers inevitably decreases economic efficiency and leads to decreased consumer satisfaction. If food retailers agree to pay a penny a pound more for tomatoes to improve the pay or working conditions for farm workers, for example, they expect to pass the cost increase on to consumers— and will likely add another penny for profits. This will raise tomato prices for consumers, including those who don't know or care about the plight of farmworkers, thus decreasing overall consumer satisfaction.

Furthermore, the willingness of some consumers to pay more for the same tomatoes is "economically irrational," since presumably there will be no tangible differences between tomatoes produced under favorable and unfavorable working conditions. This leaves the fate of farmworkers to be determined by economically irrational consumers who can afford to pay more for tomatoes. "Free choice of employment," "just and favorable conditions of work," and "remuneration ensuring … an existence worthy of human dignity"[2] are basic human rights, according to the United Nations Declaration of Human Rights—which the United States refuses to endorse. Rights are not privileges to be granted at the discretion of employers or wealthy consumers. Rights depend on social justice—not economics. Economies afford no more respect for the "rights" of workers than for the "rights" of land or capital. They are all just factors of production.

Market economies function to meet our needs as consumers, not as workers or as members of society. Whatever economic value we receive from our work is realized only by consuming or using what we buy with the money we earn from working. Whatever we sacrifice as workers must be compensated by the benefits we receive as buyers or consumers. Unfortunately, those who benefit most as consumers are rarely the same people who sacrifice most as workers. In addition, the lack of economic completion in today's market economy allows some to extract profits from the system rather than reward workers for their efficiency or pass the savings on to consumers. Publicly traded corporations, being rational economic entities, have no incentive to do anything for the benefit of workers or consumers unless it adds to their economic bottom line.

The food industry clearly has an economic incentive to minimize labor costs, regardless of who benefits and who pays. According to the U.S. Department of Agriculture, "wages, salaries, and contract labor expenses represent roughly 17 percent of total variable farm-level costs and as much as 40 percent of costs in labor-intensive crops such as fruit, vegetables, and nursery products."[3] The nonfarm sectors of the food system are even more labor-intensive, resulting in labor costs accounting for roughly fifty cents of each food dollar of U.S. consumers. So, it is naive to expect industrial farmers or food corporations to gratuitously increase the compensation of farm or food industry workers, or to willingly grant workers their basic human rights.

The fundamental problem is a failure of society to recognize the full value of work. In capitalist economics, work is considered to be inherently unpleasant or distasteful. The money gained from working is the only reward for giving up the alternative of enjoying leisure. Work would never be willingly undertaken without some offsetting economic compensation. In economic thinking, there is no recognition of any positive value of work apart from the economic value derived from the consumer market value of whatever is produced.

While people should expect to work in order to meet their basic needs, even if the economic remuneration is meager, work can also produce social and cultural value. Yet economics gives no consideration to the fact that work helps give purpose and meaning to life. The sense of dignity arising from meaningful work can translate into a sense of self-worth that goes far beyond survival or subsistence. The admiration and respect granted by fellow workers, employers, or customers for a job well done may far outweigh any additional economic compensation. Many workers actually enjoy their work. Many more undoubtedly would do so if they were afforded their basic human rights to free choice of employment, just and favorable work conditions, and remunerations sufficient to ensure an existence worthy of human dignity.

To break the bonds of economic slavery, we must value humans as multidimensional beings, not biological machines. We are social beings capable of receiving tremendous personal value from positive human relationships—even relationships that produce nothing of economic value. We are spiritual beings capable of receiving tremendous ethical value from a life of purpose—including our life of work. Work is not a burden but a privilege, at least when performed under conditions that respect our basic human rights as workers.

We are not just consumers; we are also thoughtful, caring workers and responsible members of society. Our preferences as consumers cannot be allowed to take priority over our rights as workers and global citizens. All workers, not just farmworkers and food workers, will continue to work under conditions of economic slavery until our market economy is forced by civil society to recognize and respect the full economic, social, and cultural value of work.

NOTES

1. Jane Kolodinsky, "More than One Meaning of 'Chain' in Food Chains: A Documentary Film," Journal of Agriculture, Food Systems, and Community Development 5:1 (2014), 197–98, doi: 10.5304/jafscd.2014.051.011.
2. United Nations, Universal Declaration of Human Rights, Article 23, 1948, www.un.org/en/documents/udhr.
3. U.S. Department of Agriculture, Economic Research Service, "Farm Labor," n.d. www.ers.usda.gov /topics/farm-economy/farm-labor.aspx.

Originally published as John Ikerd, "Rethinking the Value of Work," *Journal of Agriculture, Food Systems, and Community Development* 6:2 (2016), 5–7, doi:10.5304 /jafscd.2016.062.003. Copyright © 2016 by New Leaf Associates, Inc.

Timberwolves

AIMING AT BEULAH

LYN ARCHER

At the gun show, he braces the metal yoke against my forearm as we practice snapping the long rubber strap. The Wrist Rocket is intended for the family of raccoons who've been breaking into our chicken coop each night, during the hour before my housemates wake, the hour I return from work: the Devil's Hour.[1] We leaf through libertarian propaganda, fondle old war medals. He passes me a pearl-handled Derringer: *Here. For your garter.*

On Sundays there are only two or three of us working. The place is tiny, and sits on a small triangular lot adjoining a body shop. All the seats and benches are upholstered in black automotive vinyl. The dressing room curtains are blood-red velveteen. The strip club is named after the Devil.

From a back table, a man wads up single bills and pitches them at me. Sunlight shafts from the open door, across the video poker machines, the television. The regulars glance, reflecting in the paneled mirror. I twirl in triplicate. Night falls, candles light. Money turns in the ceiling fan's breath. A row of dollars line the stage's rail. My g-string becomes a wreath of laurels. After last call I stand in my stocking feet on the rubber floor mat. The bartender's hand stabilizes my wrist as we agitate the cocktail shaker, so that ice crystals form but don't melt.

The hour between night and morning is silent. The big, pink house sags like a wet cake-box. Passed between students for over a decade, it shelters a dozen or so of us in vague communality.

Vines curtain the porch. Someone peeks out from the sofa. I tiptoe over half-made protest signs, forms cocooned in sleeping bags on the floor. A family of bandits. Twelve hours ago, as I emerged from the shower in my kimono, a group of denim-and-leather-clad travelers glared at me stonily from the kitchen table. There's a stigma that sex work isn't real work, natural work, work of value, that places it in opposition to agrarian work.

Mice poke their heads between the electric coils on the stove. My housemates and I meet in the kitchen. Their days begin in the hoop house, at the market, in the row-crops, in the library—with Steiner, Berry, Fukuoka. I eat bites of their blueberry pancakes. We brew coffee, pass around a tepid Corona I snuck from the club.

Farmers and strippers hurt and heal the same. We tape our ribs, brace our wrists, ice our knees and backs. Our fates are tethered to violent forces: weather and seasons, tourism and trade, economy and law. Two draft animals of unequal power, if yoked together, are unable to plow an even furrow—one drags the other off on a tangent. Land, if exploited, eventually refuses to produce anything of value. Our bodies, too.

On our bedroom windowsill, my partner has lined up our ammunition: a row of pebbles, bottle caps, loose change. We wait for the familiar scuffle in the yard below. Offense and defense blur together. For every strip club named after Hell there's another called Heaven. Paradise. Eden. There's a dive bar nearby, named after Beulah—

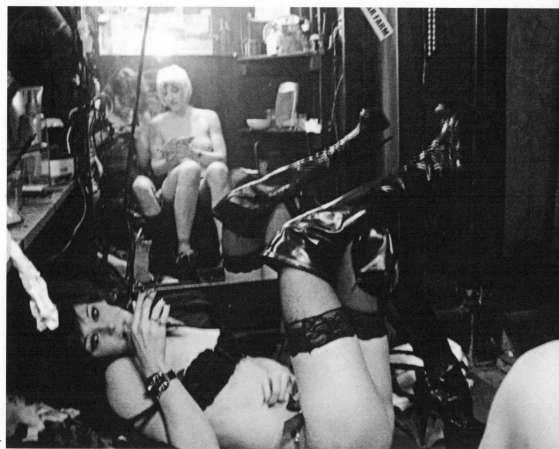

Lyn Archer

William Blake's vision of "a dreamy paradise where the sexes, though divided, blissfully converse in shameless selflessness. Beulah is available through dreams and visions to those in Ulro, the utterly fallen world."[2]

Beulah is intrinsically beyond us, yet, we aim at it. Doing so creates a state of grace, where oppositions can abide. Agrarian and sex worker, the two oldest vocations, serve life through an immediate relation between body, land, spirit. Where night and day touch, we meet.

NOTES

1. "The term Devil's Hour applies to the hour of 3 a.m. or 3:15…. The term may be used to refer to any arbitrary time of bad luck…. Women caught out late at night could have been suspected of witchcraft if they did not have a legitimate reason to be out." https://en.wikipedia.org/wiki/Witching hour.

2. Gourlay, Alexander S. "An Emergency Online Glossary of Terms, Names, and Concepts in Blake." The William Blake Archive. https://blakearchive.org.

DIFFERENTIAL CONSCIOUSNESS AND AGRARIAN EXPRESSION

RACHEL WEAVER

Today's young farmers, agrarians, and stewards often function within yet beyond the demands of the dominant ideologies of capitalism, industrialism, and individualism. Differential consciousness is a philosophical survival skill for our ambiguous future.

Chela Sandoval elaborates on oppositional consciousness in her pioneering 1991 essay "U.S. Third World Feminism: The Theory and Method of Oppositional Consciousness in the Postmodern World."[1] Oppositional consciousness helps chart the variety of realities in diverse cultural regions, critical places, and dystopian landscapes. Differential consciousness enables movement among and between the lived differences of our experiences. In "Mexicanas' Food Voice and Differential Consciousness in the San Luis Valley of Colorado," Carole Counihan defines differential consciousness as "the ability to acknowledge and operate within [demeaning and disempowering] structures and ideologies but at the same time to generate alternative beliefs and tactics that resist domination."[2]

Differential consciousness entails strength to confidently commit to well-defined, potentially dominant structures of identity for temporary distinctions of time.[3] Urban farmers often work additional jobs in order to earn income for general living expenses. We desire jobs that are complementary to our agrarian passions, but must also conform to the availability of off-farm professions. We are often part-time or contracted employees; examples include warehouse work, desk jobs, educators, custodians, and restaurant employees. With each position, we must confidently commit to the dominant structures and ideologies of our jobs while understanding that the arrangements are temporary.

Young farmers expressing differential consciousness are able to complete the daily farm tasks, prepare for evening work, and withstand nighttime commutes. We simultaneously realize our fragility in striving toward our creative and

productive potentials. Throughout the seasons, I experience severe weather, crop failures, and slow market days. Farms are hubs for experimentation, works of hospitality, and community expression. We acknowledge the human necessity to ask for help, to discern genuine advice, and to receive the assistance of others.

Differential consciousness within the young agrarian requires flexibility to self-consciously transform the identity to another oppositional tactic, according to the power formations present.[4] We are able to determine and differentiate forms of power, authority, and influence. Fieldwork creates team dynamics, influenced by cultural and seasonal climates. Market managers cope with expanding bureaucratization. Environmental educators seek to develop experiential curricula. Community organizers cultivate networks and enhance local food access. Each workday asks the agrarian to extend their scopes of perception.

Differential consciousness within the young agrarian requires "grace to recognize alliance with others committed to egalitarian social relations." Part of this alliance means revealing our vulnerabilities and understanding the complexities of the "women and men of the same psychic terrain."[5] We form restorative utopias with simple acts—remove weeds, wash greens, prep salad. We traverse the essential character of agrarianism. The psychic terroir and aesthetic competence of the young steward understands the flavors of the landscape, and the collective harmony of the soil, life, and weather.

NOTES

1. Chela Sandoval, "U.S. Third World Feminism: The Theory and Method of Oppositional Consciousness in the Postmodern World" *Genders* 10 (1991), 1–24.
2. Carole Counihan, "Mexicanas' Food Voice and Differential Consciousness in the San Luis Valley of Colorado," in *Food and Culture*, edited by Carole Counihan and Penny Van Esterik (New York: Routledge, 2008), 354–368.
3. Sandoval, "U.S. Third World Feminism."
4. Ibid.
5. Ibid.

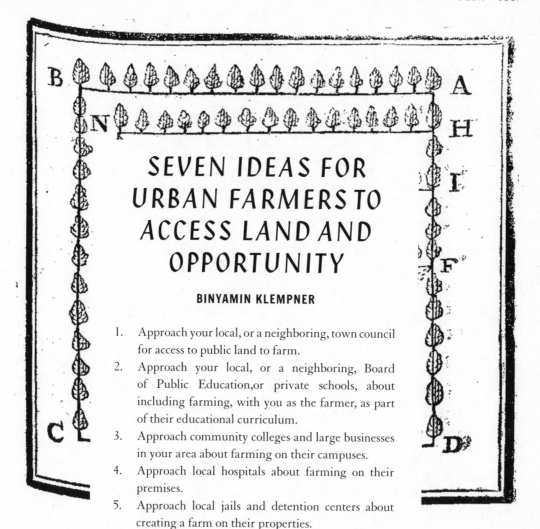

SEVEN IDEAS FOR URBAN FARMERS TO ACCESS LAND AND OPPORTUNITY

BINYAMIN KLEMPNER

1. Approach your local, or a neighboring, town council for access to public land to farm.
2. Approach your local, or a neighboring, Board of Public Education,or private schools, about including farming, with you as the farmer, as part of their educational curriculum.
3. Approach community colleges and large businesses in your area about farming on their campuses.
4. Approach local hospitals about farming on their premises.
5. Approach local jails and detention centers about creating a farm on their properties.
6. Approach the clubhouses, churches, synagogues, and mosques in your area about farming on their lands.
7. Offer the folks in your area the service of building mini farms on their lawns, rooftops, and porches.

Remember to use the available resources at hand to grow where you are.

The Farmers' Union

Entered as second-class matter August 29, 1912, at the post office at Salina, Kansas, under Act of March 3, 1879.

VOL. V.   SALINA, KANSAS, THURSDAY, OCTOBER 10, 1912   NO. 7

OFFICIAL ORGAN OF THE FARMERS' EDUCATIONAL AND CO-OPERATIVE UNION OF KANSAS

TOPICS FOR DISCUSSION

Questions Suggested by the National Convention for the Next Twelve Months

At the last national convention of the Farmers' Union the following subjects were adopted as suitable for discussion by the local Unions during the next twelve months:

Economics

Who is responsible for the high cost of living?

What effect does gambling in farm products have on the prices of the same?

How does the tobacco trust affect the price of tobacco?

Which is the best policy for public improvements, direct taxation or bond issues?

Will a parcels post benefit the farmer: if so how shall we proceed?

Does supply and demand govern the price of farm products?

Possibilities of farming.

What per cent of the taxes do the farmers pay?

Can there be an over-production of farm products?

Conservation of timber, water, power and soil.

Have the trusts been detrimental to the general public?

Civics

Citizenship: its duties and responsibilities.

What effect will Asiatic immigration have on the future of this country?

The election of United States senators by direct vote.

Initiative, referendum and recall.

Rural Problems

Benefits of good roads—how to get them.

What can be done to keep the boys and girls on the farm?

How to make the country schools attractive.

Evils of tenantry: what does the Union offer to the tenant?

Difficulties of the farmer.

The farmers' schoolhouse, Sunday school and church.

Farmers' co-operative telephones.

Sanitary conditions on the farm.

Farm exhibits at fairs.

Business

Co-operation and its results.

Farmers' warehouses, elevators: what they stand for.

Best methods of marketing crops.

Propriety of a bureau of information as to acreage of farm products, amount of live stock on the farm, estimates of normal demand for each product and the country consuming each, with cost of transportation.

Propriety of consolidating all Farmers' Union elevators under one charter and all cotton warehouses under one charter.

Good of Order

Union pledges, promises and resolutions.

What are the qualifications of fitness for local officials?

What have I done for the success of the Farmers' Union and what has it done for me?

What must I do to be an ideal member?

What is an obligation and what does it imply?

Am I big enough to lay down my individuality for the good of the order?

How to get up attractive and successful programs for local meetings, barbecues and Union rallies: how to make the local a social center.

Duties of officers and members towards each other?

How can we distinguish between the friends and the foes of the organization?

Achievements of the Union in different states.

AUGUST
INFORMATION COMMONS

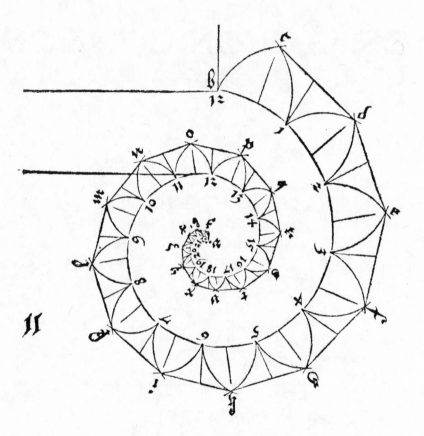

Albrecht Dürer

AN AGRICULTURALIST HIPPOCRATIC OATH

DORN COX

As agriculturalists and citizens we are not responsible for feeding the world but for creating a system by which individuals have the means to create their own abundance, feed their community, and take part in improving the system for their descendants. The promise of agriculture is not to build dependence but to create more independence, more choices, more liberty; by becoming more learned in this system, we may all become happier living in a world whose systems we collectively understand to the extent that our senses allow. We partake in this human experience by attempting to transform these systems from the invisible into the visible, by bringing what had been shrouded in darkness into the light.

If we see our actions in the context of progress and civil society as having the common purpose of building a civilization that lives on the profits of nature justly created and enhanced, then our rights and privileges associated with agriculture should also follow. It is not a battle that is won but a cumulative neverending struggle in which progress can be accelerated through generations of accumulated knowledge.

Therefore it is in our collective interests to lay out the points that unite us.

1. All members of civil society who are going to take action to improve and effect natural productive systems (agriculture) should have access to the best knowledge available on earth, and it is in our shared interest to remove barriers to sharing this knowledge.

2. All knowledge of natural systems should be accessible, inexpensive, and shared widely.

3. Knowledge should be shared so as to improve its accuracy with every generation. As tools are a reflection of available systems knowledge, we have a right to access the best tools available based on our collective knowledge.

4. Potential knowledge of our environment is infinite, but each lifecycle helps illuminate

patterns; since our knowledge of the ever changing environment also changes, all farmers and members of civil society should have the right to modify their tools to better suit their changing environmental conditions.

5. We have the right to use our collective body of knowledge to improve the productivity of natural systems, and where our labor has improved that system, the natural returns are then justly allocated based on our best understanding of science and art.

Agriculture is a human activity; therefore its application and practice are as much a product of social ecological systems and values as technical or biological limits. The cultural and public nature of agriculture, however, is often understated or misrepresented as just another business, rather than the regenerative business that forms the basis for all others. Leonardo da Vinci is credited with saying, "We know more about the movement of celestial bodies than about the soil underfoot." And that is still true. Agriculture is a fundamental human pursuit and justifies the public involvement that space exploration excites. What if were to treat each farm like a newly discovered planet, with the same public curiosity, interest, and observational, analytical, and communications tools now deployed to understand our own biosphere and our own back fields? As we understand our environmental health as a system we depend on for our wellbeing we may internalize it as an extension of the health of our physical body.

Agriculturalists have an obligation that transcends the health of the individual; the agriculturalist's actions operate on the health of a system that affects not just a single patient but the health of all life. The physician, the plumber, the banker, the bus driver, the duck, the moose, the ant, the mushroom all cycle water, carbon, and nitrogen, and all cycle through the soil that the farmer manages. Consequently, a new set of easy-to-remember principles need codifying; they should be easy to teach, and easy to hold people accountable to, just like principles taught to doctors.

It seems high time that we extend to the agriculturalist a version of the physician's oath. In its original form, the Hippocratic Oath requires a new physician to swear to uphold specific ethical standards. Of historic and traditional value, the oath is considered a rite of passage for practitioners of medicine, but why only physicians? Why not holders of public data, why not engineers, why not scientists, why not businesses, and why not agriculturalists? What about the health and well-being of our public atmosphere, our soils, our water, and our data about how the system of our biosphere functions and how we can create health or destroy it?

Because we, as agriculturalists, have the ability to contemplate our ability to improve or degrade biological systems health, we all have a public obligation to uphold the highest standards of care. I suggest that we undertake this work, with credit to Diderot's 1751 *Encyclopédie*, to demonstrate the general system to the people with whom we live, and to transmit it to the people who will come after us, so that the works of centuries past is not useless to the centuries that follow—that our descendants, by becoming more learned, may become more virtuous and happier, and that we do not die without having merited being part of the human race. I will start by taking the oath myself (adapted from Dr. Louis Lasagna's 1964 Hippocratic Oath).

I swear to fulfill, to the best of my ability and judgment, this covenant:

> I will respect the hard-won scientific gains of those farmers in whose steps I walk, and gladly share such knowledge as is mine with those who are to follow.

> I will apply, for the benefit and health of all soil, all measures that are required, and not sacrifice long-term fertility for short-term yields.

> I will remember that there is art to farming as well as science, and that planning, observation, analysis, and collaboration with other farmers may outweigh the plow or the fertilizer hopper.

> I will not be ashamed to say "I know not," nor will I fail to call in my colleagues when the skills of another are needed for the recovery of our soils.

> I will respect the privacy of other farmers, for their problems are not disclosed to me that the world may know. Most especially must I tread with care in matters of life and death. Above all, I must respect the power of nature and the power I have in daring to manipulate it for humanity's gains.

> I will remember that I do not treat a soil test or a weed but living soil, whose wellness may affect not just our own economic stability but provides the basis of life for all

future generations of humans and serves a crucial function for all life on earth. My responsibility includes these related problems, if I am to care adequately for the soil.

> I will use diversity to reduce disease whenever I can, for prevention through increased photosynthesis, planning, and systems understanding as preferable to fighting the life of the system through oversimplification.

> I will remember that I remain a member of society, with special obligations to all my fellow human beings to improve the most basic regenerative systems that support us all—those in cities and towns, and those in the country, those of sound of mind and body as well as the infirm.

> If and when I find my body unable to carry out these duties, I agree to pass along my accumulated knowledge, land, and resources to those who are willing and eager to apply their minds and bodies to uphold this oath.

> If I do not violate this oath, may I enjoy life and art, respected while I live and remembered with affection thereafter. May I always act so as to preserve the finest traditions of my calling, and may I long experience the joy of helping and improving our common soil with those who seek my help.

THE COMMONS OF POLITICAL POWER, OR WHAT FARMER IS BANANA ENOUGH TO LEAVE THE BUNCH?

THOMAS DRISCOLL

-:- NEIGHBORHOOD NOTES -:-

ers financial program. He spoke some length on cooperative marketing and the raising of better live stock and utilization of better methods of farming.

Among the witty things he said was that a farmer is something like a banana and is likely to get skinned as soon as he leaves the bunch.

O'Shea was applauded repeatedly in his declaration that the middleman had no right to take so much profit that rightfully belonged to the farmer and the consumer. He quoted figures showing the profits the farmer is losing each year.

Many men attended this meeting and prominent members of the Farmers Union from all over Linn county came to this meeting. Linn county is fortunate to have heard O'Shea.—Parker Message.

Jimmy O'Shea, known in his day as "the wild Irishman of Montana," was a charismatic secretary of the National Farmers Union (NFU) during the Great Depression. Newspaper articles documenting his efforts on behalf of NFU indicate that he was a thrilling speaker. One such item related an anecdote O'Shea shared with a meeting of the Linn County chapter of the Kansas Farmers Union attended by three hundred people on February 27, 1930: A farmer is something like a banana and is likely to get skinned as soon as he leaves the bunch.

O'Shea's witticism inspires an examination of the work of NFU today, as observed through the lens of the commons. In some ways it seems that the American farmer's ability to exercise influence on the federal government can be understood as another manifestation of the commons. It is open to the advantages and drawbacks of a resource shared among many varied parties. Elected officials and government agencies with the ability to impact our food systems and rural communities seem to have a certain amount of attention, budget, and political capital to offer the American farmer. This frank and sober assessment of farm and food politics may help those who rely in part on this particular commons for sustenance to enjoy lasting bounty instead of enduring famine and catastrophe. Everyone who produces food for other people can access policymakers by virtue of that fact, but making choices that are rational solely for the individual may degrade that resource for all in the long term.

The diversity of thought visible among the Greenhorns is worthy of celebration, though it may be challenging to navigate for purposes of national-scale advocacy. Readers of *The New Farmer's Almanac* will have different ideas about whether the federal government should be helping folks who are new to agriculture attain a foothold or whether it should just get out of the way. The production and conservation philosophies and practices they maintain are dramatically varied. But a few of the characteristics of stable commons resources are especially relevant to farmers' political power and can help participants make the best use of this resource despite their differences. These characteristics include: collective choice arrangements that facilitate the involvement of those who use the commonly held resource; monitoring of the resource by or on behalf of those who use the resource; and multiple layers of organization, with local governance over parts of the resource as the base of the pyramid building toward oversight of the entire resource.

The longevity and success of NFU, founded in 1902, stems from the organization's ability to provide these commons-management services for member farmers. It's annual grassroots policy creation and review process is an exercise in collective choice over the political endeavors of the organization. The board of directors monitors use of the organization's influence on behalf of members. NFU is a federation of state and regional organizations, many of which have organizational structures at more local and regional levels. The organization curates the civic authority of the American farmer, keeping it strong so it can be exercised in the pursuit of improving the quality of life for farmers and rural communities and enhancing the resiliency of the national and global food system.

NFU's two hundred thousand members by no stretch of the imagination can or should agree on everything, but the organization's activities help them find common ground.

Careful management of this shared political resource makes NFU a force in finding farm solutions to the dangers that lie ahead—climate challenges problems and the relentless trends of consolidation and monopolization throughout the food system. Together, NFU members helped push the Packers and Stockyards Act and Capper-Volstead Act, which serve as foundations upon which counterforces to consolidation are built. Unlike many constructions of the commons, the shared resource of agriculture's political sway is not permanently fixed in size or amount. The more that farmers can find to agree on, and the greater the variety of voices we can have sharing a common message, the bigger and richer our commons may become. That is why you should check out NFU and consider membership.

As O'Shea reminds us, we have to stick together to save our skins.

THE WONDERFUL WORLD OF NATURAL NAVIGATION

TRISTAN GOOLEY

Natural navigation is the rare art of finding your way by using nature. It consists mainly of the unusual skill of being able to determine direction without the aid of instruments and only by reference to natural clues, including the sun, the moon, the stars, the land, the sea, the weather, the plants, and the animals. It is about using your senses and then deduction.

Natural navigation should not be used instead of navigational instruments, but in a complementary way. The senses sharpen and the world comes alive when the simple question "Which way am I looking?" is answered before reaching for the map, compass, or GPS.

Once you understand the sun's arc, it is possible to use it to find direction. For example, in the United States, the sun rises in the northeast in midsummer, in the east in spring and fall, and in the southeast in midwinter. For everyone north of the Tropic of Cancer, the sun is always due south when it's highest in the sky, when is true midday. With practice it is possible to work out what it is doing at any time during the day and from anywhere in the world. Mid-morning in March? It's halfway between dawn and midday, so the sun must be close to halfway between east and due south; it will be close to southeast. A similar technique can be used with the moon, but it takes a bit of practice.

The easiest way to understand how to use the stars to find your way is remembering that if a star is over your destination, then it is pointing the right way. Most stars appear to move and so will not stay over the same place for long, but Polaris, the North Star, sits steadfastly over the North Pole and so will always point the direction for north. The stars in a part of the sky known as the celestial equator move constantly from east to west, but they always rise due east and set due west. Orion's Belt is very close to the celestial equator, and so wherever you are in the world, it can be used as an accurate guide to east or west when it is close to the horizon.

Using nature to find direction is possible without help from the sky because of the influence of the sun and weather on the ground.

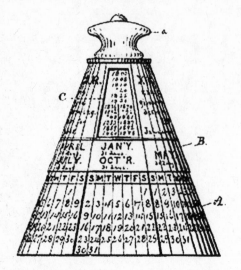

Little of what we see in nature is symmetrical, and natural navigation is often about spotting subtle differences. Trees, like all green plants, need sunlight to grow, and so their growth can be used to deduce where most of the sunlight is coming from. The "heaviest" side of isolated trees will normally be the southern side.

Mosses and lichens are quite fussy and prefer certain levels of moisture and sunlight, which means that they grow unevenly on each side of buildings and trees. Their patterns often vary from place to place, but once deciphered are consistent over large areas. North-facing roofs near where you live may have lots of moisture-loving green moss, whereas the south-facing ones may have colonies of golden lichen that are able to thrive in the sun.

It is even possible to find your way from puddles and bare earth. The sun reaches different parts of the ground more easily than others, which means that two sides of a path often reveal a clue: if one side is dry and dusty and the other is wet and muddy, then it is time to solve the puzzle of where the sun has been and then use this to find direction.

It is not just the sun that leaves footprints, but the wind too, which combs the tops of trees and each day moves clouds in a way that can be helpful. The wind can even be used just from the feel on your face and the buffeting sound in your ears.

Animals can teach us much about navigating naturally. Observing the birds' daily patterns can reveal the location of land from sea or vice versa, and watching their annual migratory patterns gave the earliest explorers clues to the location of islands in the cold North Atlantic and warmer waters of the Pacific.

There are clues in the water itself. Sailors, from the ancient Greeks through the Arabs and Vikings to modern day Pacific navigators, have learned to use the motion of the ocean swell to understand more about their location and the direction a boat is moving. The Micronesian masters of this art can tell which way their canoe is heading when they are lying down with their eyes shut.

From ancient to modern times the number of natural clues that have been used for finding direction mean that there is never a shortage of nature's compasses to look for—even in a city. Natural navigation is rich, diverse, and sometimes challenging subject, but it is one that can enrich every journey.

Originally published as Tristan Gooley, "The Natural Way of Not Getting Lost," BBC Radio 4, March 18, 2010, http://news.bbc.co.uk/today/hi/today/newsid_8568000/8568282.stm.

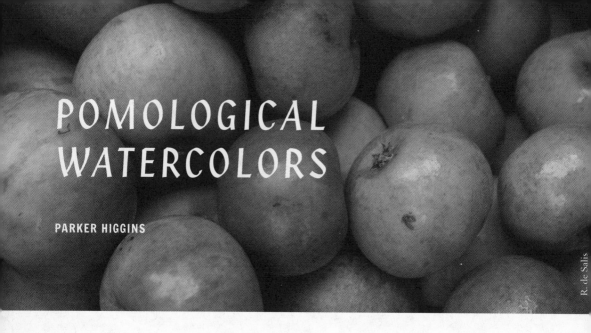

POMOLOGICAL WATERCOLORS

PARKER HIGGINS

In 1886 the U.S. Department of Agriculture (USDA) created a new Division of Pomology. Farmers faced an explosion of new and unknown fruit varieties coming from growers experimenting in their orchards and from plant explorers returning from countries around the world.

This new division hired artists to create scientifically accurate images of these fruits, to document and disseminate cutting-edge research findings. The artists drew and painted fruit over the next five decades. The resulting Pomological Watercolor Collection is massive: over 7,500 images, some half of which depict apples. They are largely painted, with a few line drawings and lithographs thrown in. While they may have been created with scientific rigor in mind, their beauty is undeniable: vivid colors and near photorealism captured by the expert hands of an artist that had already painted maybe hundreds of other varieties of apples.

But the real wonder of the collection is

taking it all in together, witnessing the breadth and variation between the fruits included—and also the difference between those century-old varieties and the supermarket fruits we know today. Many of the thousand-plus varietals of apples represented in the collection no longer exist, and in some cases the paintings are the only documentation we have.

The USDA knows the extraordinary value of this collection. A September 2000 article in the Department's *AgResearch* magazine describes it as a "priceless but little known legacy for all Americans." But this legacy for all Americans was almost inaccessible. At that time, only a vanishingly small number of people could look at this treasury of paintings. That began to change in 2010, when the National Agricultural Library embarked on a year-long digitization project, scanning and cataloging the entire collection for the first time. The library put the scanned images on the Web, but while it had created high quality scans for preservation, it put only low-resolution thumbnails online. These lower-quality images were pretty to look at, but would be unsuitable for printing or serious research.

This was the condition in which I first encountered the collection in 2015. It was beautiful and breathtaking in scope, and yet so far short of its potential. Curious, I poked around for more information through a Freedom of Information Act (FOIA) request. It was the USDA's response to my FOIA request that told me that incredibly detailed images existed somewhere out of reach. So I filed another FOIA request, this time seeking the high-quality files themselves. To its credit, the library rose to the occasion, placing the files online for anybody to access.

Of course, ensuring access is just the beginning—especially with a collection as intimidating as these watercolors. I wanted the fruits to meet people where they were, so I uploaded the entire collection to the Wikimedia Commons. I even created a Twitter bot, @pomological, that tweets a random image from the collection every three hours. In the six months since the higher-quality collection was released, I've watched its images bring people joy and inspire wonder and curiosity. This was, for me, a rare opportunity to grow the shared public domain of cultural material we can all enjoy and build upon.

That public domain—the commons—has stagnated of late because of a shortsighted legislative decision made nearly twenty years ago. The Copyright Term Extension Act of 1998—sometimes described as the "Mickey Mouse Protection Act" because of Disney's investment in its passage—at a single stroke lengthened all existing copyright terms by twenty years. The effect was that cultural works that should have

entered the public domain in 1999 now remain restricted until 2019; the free and unfettered access we should have had to works in 2000 we don't get until 2020, and so on. In other words, since 1999, we've been in a long drought without copyright expiration. For a person age thirty-five, that's half a lifetime. The cultural changes brought about by the Web have happened in a world where the public domain grows only by trickles.

A cultural commons that doesn't expand feels ancient and irrelevant. Where once people could freely build on the popular culture of their youth, or at least that of their parents, we now must look to the memories of many generations past. The consequences of that perceived irrelevance are predictable: people feel removed from the production of culture, as if it's not an arena in which they can participate. There's a striking parallel to the modern fruits of a supermarket culture, the ones that look so little like their watercolor forebears. In both culture and in agriculture, we have fostered homogeneity.

In the copyright context, a free and participatory culture can reflect the variety in each of us. By contrast, a top-down culture encourages the lowest common denominator. The Pomological Watercolor Collection is beautiful on its face, and on further examination more profoundly so. At its core, it is an artifact of an effort to share discovery, progress, and creativity with the world. A collection that is accessible to all—and that encourages everybody to expand it, to reshape it, to build upon it—is one that continues to advance that noble mission.

TREES AS EDUCATORS.

The beauty and value of trees about the home may be readily conceived if one will go from a place well protected by a grove of suitable varieties tastefully arranged to a prairie dwelling absolutely bare of arboreal adornment and shelter. It is like geing transferred from the ideal garden of Eden to the desert of Sahara—from the security of protective friends to the blackness of desolation. In one place the birds make melody morning and evening; in the other the only music is the howling or moaning wind—the nervedestroying wind—as it seeks unwelcome entrance at every crack and crevice in thecheerless abode.

The former spot is partially sheltered from the hot drying winds of summer and the Alaskan blasts of winter; the latter is at the mercy of theunfriendly elements from one years' end to the other. In one house you may expect to find some of the comforts of life and some of the cheerful signs of culture and refinement; in the other hopelessness and despair should be depicted on the countenances of its inmates. A well ordered household and happy children are the natural outgrowth of the home where the trees are loved; ill-temper, slovenly habits and discontent are to be expected in the dreary, uninviting quarters of families reared without the civilizing influences of God's sweetest teachers, trees.

The lessons they silently inculcate are like the never forgotten benedictions of a happy childhood. They live in memory when age creeps on apace, and exert a wholesome influence on all who are fortunate enough to be reared within the raceh of their gracious presence.

The trunks are suggestive of strength, grandeur, sublimity and endurance.

From the branches we may learn the lessons of dependence and submissiveness. The nourishment they receive from their fountain of sustenance remind us that none are above the helpful ministrations of loyal friends. They also gracefully bend before the sudden gale, but seldom break.

The leaves teach the beautiful lesson of service. They gather up the sunshine and shower and theunobserved gases from the atmosphere to coax the parent tree to store its cells against the time of need; they hold a dainty parasol over thebirds and beasts that seek shelter from the burning rays of the midsummer sun; and in autumn they die and drop to the ground for the benefit of the branches and trunk which bore them and theroots which gave them nourishment.

Trees teach symmetry and beauty sa well as therugged virtue of self-reliance and thelofty sentiment of living for others.

To one who has been brought up under their sweet and helpful influences a treeless home is an abomination—a misnomer.

HON. EUGENE SECOR.
Of Stae Board of Horticulture. Forest City, Iowa.

B 1714 Big Trees, Felton, Santa Cruz Co.   Photo., San Francisco

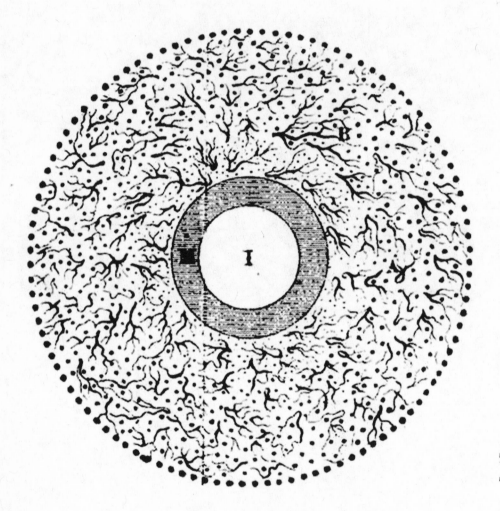

René Descartes

MYCORRHIZAS

DYLAN SMITH

Mycorrhizas are hybrid organs of molecular exchange that exist within and around the roots of plants. They are both fungal and plant, and not exclusively one or the other. Fungal hyphae (the vegetative body of a fungus) wrap around—or in some cases even penetrate into—the root cortical cells, and travel outward deep into the surrounding soil. At the edges of its body, the fungal component sequesters water and mineral nutrients like nitrogen and phosphorous from the soil, transferring them through its own cells directly into the plant. In return, the plant funnels photosynthesized carbon in the form of sugars down from its leaves into the fungus.

Mycorrhizas are described as providing extended root systems for plants. But in a sense, mycorrhizal fungi are the plants' root systems; at least 80 percent of plant families form mycorrhizas, and almost all of them are incapable of living without their fungal partners. Similarly most mycorrhizal fungi cannot degrade cellulose or lignin—the chief sources of carbon in soil—and thus are almost entirely dependent on plants for their food. Mycorrhizas are the most common mutualism in the terrestrial world, and their origins are ancient: the oldest known fossils of plants have mycorrhiza-like structures on their roots. It is now a widely accepted theory that plants could spread across a landscape only after the evolution of the mycorrhizal symbiosis.

In reality, this symbiosis creates a vast exchange of molecular commodities among many organisms throughout the soil commons. Fungi show variable levels of specificity toward the types of plants they will colonize, but most mycorrhizal plants are thought to be wide generalists. At least forty known mycorrhizal species regularly form mycorrhizas with pines, for example, including many of the most recognizable fruiting fungi of the forest floor. And just as common as a single plant harboring multiple microbial symbionts at once, it is increasingly being recognized that a single fungus can be connected to multiple plant partners at once as well.

Connectivity of plant species by way of mycorrhizal fungi opens a veritable treasure chest

237

of questions and possibilities about aboveground ecology and the soil ecosystem. The transfer of nutrients and sugars from different species— sometimes even phyla—of plants at different life stages and variable locations within a community invariably means that the growth of a temperate ecosystem is a collective experience. The idea that older mature trees are capable of "nursing" young nearby seedlings, and likely often do so through the allocation of carbon and other nutrients, is now commonplace.

The study of ecological successional dynamics now fully recognizes the primary colonization roles of mycorrhizal fungi within new or recently disturbed environments before the more recognizable aboveground plant landscapes begin to reappear. Consider a major ecological disturbance like fire. Many plants, especially from the family Ericaceae, maintain belowground root systems even after a high-intensity aboveground blaze. These roots harbor huge numbers of common mycorrhizal fungal species, thus maintaining a "legacy inoculum" of valuable mycorrhizal fungi that help mixed forests regenerate after fire. Even in the absence of intact root systems, fungi can remain remarkably persistent within soil through vegetative hyphae or long-lasting spore banks. Studies in the California Rim Fire, one of the largest recorded fires in the Sierra Nevada, have begun to identify typical fire pioneer species of fungi that survive and thrive in recently burned environments, already present in high abundance just years after a 100 percent crown death burn.

Adding to these complex dynamics, spatial patterns of fungi and bacteria, long thought to just "be everywhere," are increasingly being understood in greater detail. Evidence now points to the fact that fungi are not simply everywhere, but are subject to the same types of dispersal limitations and spatial biogeography that ecologists have long known about plants. Given the ubiquity of the mycorrhizal symbiosis in terrestrial ecosystems, it's no wonder that ecologists now consider plant biogeography and fungal biogeography intricately linked. Examples in New Zealand and the southern South American continent highlight the largely failed attempts at introducing nonnative pine, with experiments showing that the introduced pines associate poorly with the native fungi in these regions. Cointroduction of pine's native fungi, mainly Rhizopogon and Suillus species, have led to significantly higher rates of establishment and survivorship in the invading plants.

Ecologists talk of the era of the "black box" of belowground community dynamics. What does the added complexity of mycorrhizal and other microbial symbioses mean for our collective understanding of the evolution of life on this planet? Where does it leave us now that the black box has been smashed open? Why do symbioses remain evolutionarily stable relationships between partners that, for all we can understand, lack consciousness? Could we be wrong about this consciousness? These are questions that surely humble us before the awe of the natural world. These are questions worth asking, but that do not beg an answer.

❧ ❧

A LECTURE DELIVERED AT NEWCASTLE-ON-TYNE, 1775

THOMAS SPENCE

A lecture read at the Philosophical Society in Newcastle on November 8, 1775, for printing of which the society did the author the honor to expel him.

Mr. President, it being my turn to lecture, I beg to give some thoughts on this important question, viz.: Whether mankind in society reap all the advantages from their natural and equal rights of

Pieter Bruegel

❧ ❧

property in land and liberty, which in that state they possibly may and ought to expect? And as I hope you, Mr. President and the good company here, are sincere friends to truth, I am under no apprehensions of giving offence by defending her cause with freedom.

That property in land and liberty among men in a state of nature ought to be equal, few, one would be fain to hope, would be foolish enough to deny. Therefore, taking this to be granted, the country of any people, in a native state, is properly their common, in which each of them has an equal property, with free liberty to sustain himself and family with the animals, fruits and other products thereof. Thus such a people reap jointly the whole advantages of their country, or neighborhood, without having their right in so doing called in question by any, not even by the most selfish and corrupt. For upon what must they live if not upon the productions of the country in which they reside? Surely, to deny them that right is in effect denying them a right to live. Well, methinks some are now ready to say, but is it lawful, reasonable, and just, for this people to sell, or make a present even, of the whole of their country, or common, to whom they will, to be held by them and their heirs forever?

To this I answer, if their posterity require no grosser materials to live and move upon than air, it would certainly be very ill-natured to dispute their right of parting, for what of their own, their posterity would never have occasion for; but if their posterity cannot live but as grossly as they do, the same gross materials must be left them to live upon. For the right to deprive anything of the means of living, supposes a right to deprive it of

life; and this right ancestors are not supposed to have over their posterity.

Hence it is plain that the land or earth, in any country or neighborhood, with everything in or on the same, or pertaining thereto, belongs at all times to the living inhabitants of the said country or neighborhood in an equal manner. For, as I said before, there is no living but on land and its productions, consequently, what we cannot live without we have the same property in as our lives.

Now as society ought properly to be nothing but a mutual agreement among the inhabitants of a country to maintain the natural rights and privileges of one another against all opposers, whether foreign or domestic, it would lead one to expect to find those rights and privileges no further infringed upon among men pretending to be in that state, than necessity absolutely required. I say again, it would lead one to think so. But I am afraid whoever does will be mightily mistaken. However, as the truth here is of much importance to be known, let it be boldly fought out; in order to which it may not be improper to trace the present method of holding land among men in society from its original.

If we look back to the origin of the present nations, we shall see that the land, with all its appurtenances, was claimed by a few, and divided among themselves, in as assured a manner as if they had manufactured it and it had been the work of their own hands; and by being unquestioned, or not called to an account for such usurpations and unjust claims, they fell into a habit of thinking, or, which is the same thing to the rest of mankind, of acting as if the earth was made for or by them, and did not scruple to call it their own property, which they might dispose of without regard to any other living creature in the universe. Accordingly they did so; and no man, more than any other creature, could claim a right to so much as a blade of grass, or a nut or an acorn, a fish or a fowl, or any natural production whatever, though to save his life, without the permission of the pretended proprietor; and not a foot of land, water, rock, or heath but was claimed by one or other of those lords; so that all things, men as well as other creatures who lived, were obliged to owe their lives to some or other's property, consequently they like the creatures were claimed, and, certainly as properly as the wood herbs, etc., that were nourished by the soil. And so we find, that whether they lived, multiplied, worked or fought, it was all for their respective lords; and they, God bless them, most graciously accented of all as their due. For by granting the means of life, they granted the life itself; and of course, they thought they had a right to all the services and advantages that the life or death of the creatures they gave life to could yield.

Thus the title of gods seems suitable enough to such great beings; nor is it to be wondered at that no services could be thought too great by poor dependent needy wretches to such mighty and all-sufficient lords, in whom they seemed to live and move and have their being. Thus were the first landholders usurpers and tyrants; and all who have since possessed their lands, have done so by right of inheritance, purchase, etc., from them; and the present proprietors, like their predecessors, are proud to own it; and like them, too, they exclude all others from the least pretense to their respective properties. And any

one of them still can, by laws of their own making, oblige every living creature to remove off his property (which, to the great distress of mankind, is too often put in execution); so of consequence were all the landholders to be of one mind, and determined to take their properties into their own hands, all the rest of mankind might go to heaven if they would, for there would be no place found for them here. Thus men may not live in any part of this world, not even where they are born, but as strangers, and by the permission of the pretender to the property thereof; which permission is, for the most part, paid extravagantly for, though many people are so straitened to pay the present demands, that it is believed if they hold on, there will be few to grant the favor to. And those land-makers, as we shall call them, justify all this by the practice of other manufacturers, who take all they can get for the products of their hands; and because that everyone ought to live by his business as well as he can, and consequently so ought the land-makers. Now, having before supposed it both proved and allowed, that mankind have as equal and just a property in land as they have in liberty, air, or the light and heat of the sun, and having also considered upon what hard conditions they enjoy those common gifts of nature, it is plain they are far from reaping all the advantages from them which they may and ought to expect.

But lest it should be said that a system whereby they may reap more advantages consistent with the nature of society cannot be proposed, I will attempt to show the outlines of such a plan.

Let it be supposed, then, that the whole people in some country, after much reasoning and deliberation, should conclude that every man has an equal property in the land in the neighborhood where he resides. They therefore resolve that if they live in society together, it shall only be with a view that everyone may reap all the benefits from their natural rights and privileges possible.

Therefore a day appointed on which the inhabitants of each parish meet, in their respective parishes, to take their long-lost rights into possession, and to form themselves into corporations. So then each parish becomes a corporation, and all men who are inhabitants become members or burghers. The land, with all that appertains to it, is in every parish made the property of the corporation or parish, with as ample power to let, repair, or alter all or any part thereof as a lord of the manor enjoys over his lands, houses, etc.; but the power of alienating the least morsel, in any manner, from the parish either at this or any time hereafter is denied. For it is solemnly agreed to, by the whole nation, that a parish that shall either sell or give away any part of its landed property, shall be looked upon with as much horror and detestation, and used by them as if they had sold all their children to be slaves, or massacred them with their own hands. Thus are there no more nor other lands in the whole country than the parishes; and each of them is sovereign lord of its own territories.

Then you may behold the rent which the people have paid into the parish treasuries, employed by each parish in paying the government its share of the sum which the Parliament or National Congress at any time grants; in maintaining and relieving its own poor, and people out of work; in paying the necessary officers their salaries; in building, repairing, and adorning its

houses, bridges, and other structures; in making and maintaining convenient and delightful streets, highways, and passages both for foot and carriages; in making and maintaining canals and other conveniences for trade and navigation; in planting and taking in waste grounds; in providing and keeping up a magazine of ammunition, and all sorts of arms sufficient for all its inhabitants in case of danger from enemies; in premiums for the encouragement of agriculture, or anything else thought worthy of encouragement; and, in a word, in doing whatever the people think proper; and not, as formerly, to support and spread luxury, pride, and all manner of vice. As for corruption in elections, it has now no being or effect among them; all affairs to be determined by voting, either in a full meeting of a parish, its committees, or in the house of representatives, are done by balloting, so that votings or elections among them occasion no animosities, for none need to let another know for which side he votes; all that can be done, therefore, in order to gain a majority of votes for anything, is to make it appear in the best light possibly by speaking or writing. Among them government does not meddle in every trifle; but on the contrary, allows each parish the power of putting the laws in force in all cases, and does not interfere but when they act manifestly to the prejudice of society and the rights and liberties of mankind, as established in their glorious constitution and laws. For the judgment of a parish may be as much depended upon as that of a House of Lords, because they have as little to fear from speaking or voting according to truth as they.

A certain number of neighboring parishes, as those in a town or county, have each an equal

R. de Salis

vote in the election of persons to represent them in Parliament, Senate, or Congress; and each of them pays equally towards their maintenance. They are chosen thus: all the candidates are proposed in every parish on the same day, when the election by balloting immediately proceeds in all the parishes at once, to prevent too great a concourse at one place; and they who are found to have a majority, on a proper survey of the several poll-books, are acknowledged to be their representatives.

A man by dwelling a whole year in any parish becomes a parishioner or member of its corporation; and retains that privilege till he lives a full year in some other, when he becomes a member in that parish, and immediately loses all his right to the former for ever, unless he choose to go back and recover it by dwelling again a full year there. Thus none can be a member of two parishes at once, and yet a man is always member of one though he move ever so oft.

If in any parish should be dwelling strangers from foreign nations, or people from distant countries who by sickness or other casualties should become so necessitous as to require relief before they have acquired a settlement by dwelling a full year therein; then this parish, as if it were their proper settlement, immediately takes them under its humane protection, and the expenses thus incurred by any parish in providing those not properly their own poor being taken account of, is discounted by the Exchequer out of the first payment made to the state. Thus poor strangers, being the poor of the state, are not looked upon with an envious eye lest they should become burthensome, neither are the poor harassed about in the extremity of distress, and perhaps in a dying condition, to justify the litigiousness of the parishes.

All the men in every parish, at times of their own choosing, repair together to a field for that purpose, with their officers, arms, banners, and all sorts of martial music, in order to learn or retain the complete art of war; there they become soldiers. Yet not to molest their neighbors unprovoked, but to be able to defend what none have a right to dispute their title to the enjoyment of; and woe be to them who occasion them to do this, they would use them worse than highwaymen or pirates if they got them in their power.

There is no army kept in pay among them in times of peace, as all have property alike to defend, they are alike ready to run to arms when their country is in danger; and when an army is to be sent abroad, it is soon raised, of ready trained soldiers, either as volunteers or by casting lots in each parish for so many men.

Besides, as each man has a vote in all the affairs of his parish, and for his own sake must wish well to the public, the land is let in very small farms, which makes employment for a greater number of hands, and makes more victualing of all kinds be raised.

There are no tolls or taxes of any kind paid among them by native or foreigner, but the aforesaid rent which every person pays to the parish, according to the quantity, quality, and conveniences of the land, housing, etc., which he occupies in it. The government, poor, roads, etc. etc., as said before, are all maintained by the parishes with the rent; on which account all wares, manufactures, allowable trade employments or actions are entirely duty free. Freedom to do anything whatever cannot there be bought; a thing is either entirely prohibited, as theft or murder; or entirely free to everyone without tax or price, and the rents are still not so high, notwithstanding all that is done with them, as they were formerly for only the maintenance of a few haughty,

unthankful landlords. For the government, which may be said to be the greatest mouth, having neither excisemen, customhouse men, collectors, army, pensioners, bribery, nor such like ruination vermin to maintain, is soon satisfied, and moreover there are no more persons employed in offices, either about the government or parishes, than are absolutely necessary; and their salaries are but just sufficient to maintain them suitably to their offices. And, as to the other charges, they are but trifles, and might be increased or diminished at pleasure.

But though the rent, which includes all public burden, were obliged to be somewhat raised, what then? All nations have a devouring landed interest to support besides those necessary expenses of the public; and they might be raised very high indeed before their burden would be as heavy as that of their neighbors, who pay rent and taxes too. And it surely would be the same for a person in any country to pay for instance an increase of rent if required, as to pay the same sum by little and little on everything he gets. It would certainly save him a great deal of trouble and inconvenience and government much expense.

But what makes this prospect yet more glowing is that after this empire of right and reason is thus established, it will stand forever. Force and corruption attempting its downfall shall equally be baffled, and all other nations, struck with wonder and admiration at its happiness and stability, shall follow the example; and thus the whole earth shall at last be happy and live like brethren.

From Thomas Spence, *Pig's Meat* (London, 1795).

SEPTEMBER

SUPPLY CHAINS
AND TRADE ROUTES

CRYPTOCAPRA

ANTISTASIA

Welcome to Cryptocapra, a modest, private, experimental, underground exchange for people who want real, local, living food, and the stories behind it.

Since I have a lot of new people buying cheese and pork from me, I wanted to take the time to properly introduce Cryptocapra. If you intend to enjoy the spoils of my labor, please do me the favor of reading this so that you can understand and appreciate what goes into it, the need to keep it discreet, and the direction I am trying to take things. Just a disclaimer: this is long, and there is some gloom in the introduction, but it is gloom that we can do something about, and I am proposing a way—all sunshine, rainbows, and capra-unicorns, I swear.

In the paragraphs that follow from there, you will find information about legal issues concerning raw milk and bootleg cheese, the land, the goats, my husbandry practices, milk handling, my cheese-making, how best to store and enjoy the cheese, and last but not least, my plans to raise more pigs for pork.

So lucky that I discovered this *How to Do It Encyclopedia* when I was cleaning out the shed!

Want more details? Come out here and see for yourself. Pick my brain, get dirty, kiss a goat, and slap a pig on the ass. That's what eating local is all about! Want less? The headings below let you scroll through.

INTRODUCTION

I am offering real food, and the real story of where it comes from, which might not always be the story that you expect or want to hear. My business style is peculiar, and downright contrary, but given that I am operating alone at the far outer edges of marginal local agriculture, this shouldn't be terribly surprising. Please accept my honesty about the realities of being a food producer as a compliment to your intelligence, an acknowledgment of your devotion to strengthening our food-shed, and an invitation to further query and meaningful dialogue. I see this as a way to initiate and inspire much-needed change.

Right now there is incredible enthusiasm for local, heirloom, and traditional foods, and Baja Arizona has claimed a title at the forefront of this movement with Tucson's designation as

249

America's first UNESCO City of Gastronomy. The question that I have is this: Where is all of the local food that everyone is celebrating and demanding going to come from?

I will readily admit that there have been times when cracking open the latest issue of *Edible Baja Arizona Magazine* has felt like a slap in the face to me. The images are just so incongruous with what I am seeing and experiencing in the farming community where I live and work. Don't get me wrong, I am just as excited as anyone else to see our food, culture, and so many of my close friends getting the praise and the attention that they have long deserved, and I too have benefited greatly in this atmosphere of well-cultivated local food enthusiasm, but there is almost always a big piece missing from the story. You can create desire, demand, and markets for food with alluring hype, but you can't grow food with it. To grow food you need real resources, like water, land, soil, and people who are willing and able to do the work. We are very limited in those resources, and mostly do a terrible job of protecting and honoring what we do have.

There is a fatal disconnect that we need to come to terms with. We are imposing human-constructed social, economic, and political systems onto strained natural systems, and it does not really work. Trying to produce food in a way that harmonizes with such disparate factors is an incredible challenge, and people who eat food (yes, that means everyone) need to understand this. I place particularly high expectations for understanding on those eaters of food that demand high-quality, clean, healthy, "sustainably" and locally produced food. Yes, I am addressing you, dear reader. You need to understand what it is that you are really asking for, and you need to realize that it might not be immediately obtainable, especially where we live.

To say that the agriculture that you want is largely not being researched, subsidized, and supported by the dominant agricultural paradigm is a gross understatement. I am in awe of the size and scope of the obstacles that I have been made aware of and had to face as a beginning food producer. Need I remind you that the cheese business many of you have been faithfully supporting for the past two years is illegal? I didn't set out to do this, but I needed to do something to make a living for myself out here. I was never supposed to try, much less succeed, and now I have really hemmed myself into a corner. I can't grow Cryptocapra. I can't promote myself or use my business experience as leverage in a lot of situations, and I can't just keep on doing this indefinitely because the pay sucks and the amount of work and risk I have taken on as a single person is totally unreasonable. Cryptocapra needs to evolve into and become a part of something else, and I have a vision of what that should be.

Your food choices really do matter. I would like to propose a way to make them actually count for something, and give people who want to eat local food an opportunity to invest on a deeper level and have a stake in this dirty game. The enthusiasm, talk, and ideas need to be put into practice on the ground to build the local agricultural system that people want. I am proposing that part of that effort should happen right here at my dreamy ranchito in Elfrida, a hundred miles from Tucson, where there is agricultural infrastructure and affordable land, and it can be scaled to a level that is highly productive and economically viable.

My project involves the purchase or lease of the property adjacent to my land, consisting of an irrigated quarter section with a center pivot, and a small well-equipped meat processing facility. I envision the field being put to use for experimenting with and growing a variety of lower-water-use, higher-value heirloom crops as well as pasture and feed that could be used to raise animals for processing at the shop, with zero miles traveled to the slaughter facility. Ultimately, I would like to develop a mixed-species, mixed-use, low-input, integrated system that produces high-quality, high-protein, high-calorie staple foods for local markets. I want to involve a wide variety of partners and see it serve as a vehicle for research, experimentation, integration, and the development of much needed marketing and distribution networks that can support other local farms and ranches.

KEEP IT CRYPTO

Cryptocapra comes from the Latin *crypto*, meaning "secret" or "hidden," and *capra*, which means "goat." Why crypto? Because it is illegal, that's why. The food and dairy industry is very intensely regulated in ways that are particularly disadvantageous for small-scale and artisanal producers.

It is legal to sell raw milk in Arizona under certain conditions, and it is even possible to have a state-licensed raw-milk dairy. If, however, one is selling a value-added product, it is supposed to be both a licensed dairy and a licensed food producer. I am not either of those things, nor could I be under current circumstances. Truth be told, if I had all of the help and the money that I needed to set up a licensed dairy, I still wouldn't want do it, because a dairy like that requires the use of enormous amounts of water, energy, high-cost materials, and chemicals in order to maintain state-approved levels of sanitation. By its very nature it has to be large-scale, consumptive, and paperwork-intensive. None of that appeals to me. I am operating in a low-cost, low-tech way with the resources that are available to me, and practicing the type of holistic goat husbandry that I believe in. There is plenty of room for improvement, but I manage.

I am offering a very finite amount of real food. A lot goes into it. If you want what I have to offer, appreciate what I am doing, and are interested in supporting me, then I am of the opinion that we should have the right to engage in a modest private exchange to our mutual benefit. Please honor, protect, and respect this by keeping it discreet. Please don't assume that other producers and local food enthusiasts will necessarily approve of what I am doing. There are people who are actually hell-bent on shutting down unlicensed small producers. Keep the crypto in Cryptocapra!

THE RANCHITO

The ranchito is located on 160 acres of mesquite-invaded rangeland in middle of the Sulphur Springs Valley, a hundred miles from Tucson. My property does not have irrigation rights, but much of the surrounding land does. This range is good goat country, even in its somewhat degraded state. I purchased the property at the end of 2012 and moved here in August 2013. It had been nearly destroyed and then abandoned by the previous owner, and I am still in the process of salvaging and fixing things.

The goats are purebred Alpines, and the does are heavy producers. I have both males and females but am not currently keeping a breeding buck. Two generations (eight goats) have been born here and were raised by their mothers on the range. The goats are free to come and go from the barn and the corral to browse during the day, as long as I am around (I usually am). They don't typically go too far from the homestead, which lets me keep an eye on them. I herd them out to use the forage that is farther out on the land.

These are not pen goats. Free-ranging has manifold benefits for the animals and their health. It also means that less arable irrigated land is being used to support them. Seasonal differences in their diet result in seasonal differences in the flavor and quality of the cheese. Predators are kept at bay by my two livestock guardian dogs.

DIET AND SUPPLEMENTAL FEED

The bulk of their diet comes from the wild available forage on the land. When I can't let them out, like on town days, they are fed locally grown hay. The lactating females are supplemented with alfalfa and an oat–sunflower seed mix on the milk stand. They all have access to plain and mineral salt.

MILK AND MILK HANDLING

Right now I am milking four does in various stages of lactation. I milk by hand into a stainless-steel milk pail and then filter it into large stainless canisters or glass jars. I make cheese in four-gallon batches, so as soon as I have four gallons, I process it.

All of my milking and cheese-making equipment are professional grade and used exclusively for that purpose. I sterilize everything either with bleach or with steam on a regular basis. I milk on my back porch, make the cheese

in my kitchen, and generally do whatever I have to do to make it work. Hand milking means that I pay intimate attention to my girls every day, and it allows me to pick up on any kind of problems or issues they might be having right away. I keep daily milking records and notes for each of them. Before I started Cryptocapra in spring 2014, I sent milk samples off to the lab for testing, and everything came back clean and good. Since I have had no health problems in my herd whatsoever, I have not bothered retesting. All of the new milkers that I have were born here.

RAW MILK

Yes, it is controversial. You may or may not already have an opinion about it, but I believe that if you are going to choose to consume it, you should educate yourself. If you do some research you will easily find well-backed arguments for and against consuming raw milk. It is your responsibility to decide what is the right choice for you. Keep in mind that foodborne pathogens can and do end up in all kinds of foods whether they are pasteurized and produced under sanitary or sterile conditions or not.

Cryptocapra cheese is made from raw milk almost all of the time. As a rule, I typically do not pasteurize. However, for the past two summers, after the rains, I have had a problem with wild (and harmless) yeast ruining my cheese, and so I have had to pasteurize for those couple of months both years.

THE CHEESE: A PRODUCT OF THE SEASONS

Cheese-making is one of those things where the slightest change in variables will often produce a dramatically different result. My cheese turns

out the way it does because of the particular ways that I do things, and it changes with the seasons. Doing everything by hand is time- and labor-intensive, and therefore it is also self-limiting in terms of scale.

I am also limited by environmental constraints. I work with the natural conditions and as few extraneous inputs as possible. I don't fight with the weather, and so I don't make the same types of cheese all year. Instead, I am constantly adjusting my cheese-making to suit the ambient temperature and humidity. At this point, I have things figured out well enough to be getting consistent results. Still, it is entirely normal and expected for truly seasonal artisanal cheese to have variation in its flavor, texture, and moisture content, and mine most certainly does.

› Chèvre is a very basic, soft, lactic-fermented goat cheese. It is so simple and mild that it cannot obscure anything, so it seems to express the flavors and textures of the moment more than any other cheese. I offer plain chèvre year-round, and a variety of flavored chèvres during the holiday season. It freezes beautifully, and will keep in the fridge for at least ten days in a sealed jar or container. Chèvre is the most flavorful and spreadable when served at room temperature.

› Feta is a great cheese for our desert climate because it doesn't mind the heat. It is an aged rennet-set cheese made using a mesophilic culture. Like all feta, mine is salty, but not excessively so. It keeps for months packed in brine in the fridge. It's available year-round

> The three moldies are different versions of aged, mold-ripened chèvres. I only make them in the cool-season months, so they are generally available from Thanksgiving through March. They range from mild to very strong in flavor, depending on their age. It takes about a week of draining, salting, and drying before they go into the cheese cave, and I generally consider them to be in the early stages of their prime about three weeks after that. They are very much alive and are actively changing on a daily basis as the bacteria, molds, and milk work their magic under the bloomy rind. Each individual cheese is distilled out of about a quart of milk.

> Grassland Crottin is the Cryptocapra version of the traditional crottin from the Loire Valley of France. This cheese can be aged for a really long time, to the point where I call it a "fossil." I have found that strong-flavored fossilized crottin pairs really well with IPA.

> Rangefire Ash is made in a popular style in which ash, activated charcoal in this case, is used on the surface to neutralize the pH and aid in mold development.

> Capra Cloud is the product of my own tinkering.

> Hard cheeses: I have been experimenting and having great success with a variety of hard cheeses, but I am not currently able to produce it on a scale that is economical.

Because of this, I have not been offering it for sale, but lately some people have expressed a willingness to pay the true cost. Each cheese is made with four gallons of milk and weighs between four and five pounds waxed. If you like it enough that you are willing to pay me for the milk, the culture, a whole day of work, and all the piddling around during the two-plus months of aging, then yes, you are free to commission a custom-made hard cheese from me. What is the cheese cave? Mine is an old upright freezer with a dampened towel hanging inside to maintain humidity. I do not have it plugged in. When I can maintain a temperature around fifty-some degrees in the spare room by the appropriately timed opening and shutting of windows and doors, mold-ripened cheese season is on. It usually lasts from October through March.

> Cajeta, a.k.a. dulce de leche or caramel, is made with only two ingredients: milk and sugar. Cajeta over vanilla ice cream with fresh local pecans and a pinch of sea salt has become a favorite around here, but that is just one of many ways to enjoy it. It is wonderful on fresh or baked apples and makes just about any dessert pastry insanely indulgent and good.

PORK

I am currently raising my second batch of pigs. The pork from the first batch turned out really well. Integrating pigs has been a really important way for me to make the best use of the whey and mistakes from cheese-making, and to increase my income. My pigs are very spoiled, and very happy. I play with them every day, and yes, they have names.

DIET AND FEED

I am feeding a diet from local sources, which are ample and reasonably priced, but not necessarily ideal. Right now I am feeding a complete grower feed from Maid Rite in Willcox, alfalfa (there's no GMO alfalfa in this area so far), and a mix of milo sorghum and corn that is soaked in whey. They also drink a lot of whey and get plenty of bolting veggies and scraps from the garden. Now that they are mature and tame enough, I will start taking them for walks with the goats on the range where they can get lots of exercise and root around for whatever pleases them. They seem to

like young tumbleweed, which is great.

There are zero sources for local organic feed out here, and now there are zero sources of local non-GMO corn. The one farmer who grew it has sold out to the new mega-dairy and is gone. I have asked everyone I know, and they all have the same answer. If there are any farmers growing it, they are keeping it for their own use. This is why I am adding sorghum to their diet in a ratio that replaces half the corn. It uses less water and is a non-GMO feed crop. I am not going to abandon feeding them the locally available feed corn altogether, because it is the only cost-effective way to raise hogs, and corn finishing is important for the quality of the fat. If I didn't live in a place where there was local feed growing that I could

Feeding time.
Copyright 1905 by Martin Post Card Co.

buy inexpensively by the ton, I can't see how it would make any sense to raise pork for sale at all.

The hard reality is that over 95 percent of the corn, soy, and cotton grown in this country is GMO. Organic feed is being imported from other countries like Canada, China, Romania, and India in order to produce most of the organic poultry, eggs, and pork that you see at the store. It comes from far away, and it is crazy expensive, and I am not going to bypass the local agricultural economy in order to procure it and sell pork to purists for five times the price. This is the current reality of local agriculture.

I wish I could offer you a crispy, smoky, salty, delicious piece of bacon to help you swallow these bitter truths, but I am all out until the next butchering.

PROCESSING

The options have become even fewer because Willcox Meat Packing gave up their state inspection license in September, leaving Guzman's and U of A as the only options for producers in southern Arizona who are direct marketing their meat to customers. There is, however, an available loophole in the system that I plan to exploit for the foreseeable future. I can sell a customer a whole animal, and then they can have it processed in their name. The price will be calculated by the live animal weight. I will deliver the live animal and your cutting and processing instructions to Willcox, and it will be processed "custom exempt" in your name. That means that every package is stamped "not for sale." You pay Willcox for the processing, and you can either pick it up or have it delivered; the last time I checked, they were delivering to Tucson.

Yes, that is a lot of pig. The average carcass weight of the last five was around 170 pounds, but more than one person or family can share a pig, and I will be happy to facilitate this process to a certain degree. I will guide everyone through filling out the cutting instructions, and we can team people up on a hogs in a way that gets the best use out of each animal and makes people as happy as reasonably possible. It is somewhat more complicated, but I think it is doable. Unless something changes, this is where things are headed as far as access to local meat goes. It also means that you will have an opportunity to expand your culinary horizons as you find ways to use the whole animal.

NITRATE AND BLOODSHED IN SOUTH AMERICA?

RYAN BATJIAKA

Nitrogen fertilizer used to be really hard to come by. If farmers needed to fertilize their fields with N, they basically had two options: Plant legumes that use bacteria to elegantly pull nitrogen out of the air, or apply manure. While these were important strategies, yields were often held back by low levels of nitrogen. This all changed in 1909 when Fritz Haber found a way to synthetically produce ammonia. But before this discovery, nitrogen limited food production in many parts of the world.

As farmers recognized, manure was an important source of nutrients for their fields, and bird manure, or guano, was no exception. The word *guano* actually means "fertilizer" or "manure," from the Quechua *wanu*. Bird guano was an even more potent source of soil nutrients than the manure European farmers were used to, with an NPK up to 16-12-3. That's pretty high; cow manure has an NPK more like 0.75-0.25-0.5. In a world where everyone wanted nitrogen for their farms (and for gunpowder), bird guano was an extremely valuable commodity.

So imagine finding literal mountains of the stuff. That's what you get when you have thousands upon thousands of birds excreting their meals of seafood onto islands that receive little rainfall. The Chinchas Islands off the coast of Peru met these requirements and had heaps of precious precious guano layered 150 feet deep.

Pictures of the Chinchas Islands in the 1860s show dozens of anchored ships and a whole lot of guano. The mountains of precious poop were excavated and shipped off to Europe and the

Americas. Peru started raking in the dough, and fortunes were made; chemical conglomerate Grace got its start here. Guano islands were such an important resource that in 1865, the U.S. Congress passed the Guano Islands Act, which allowed citizens to claim any unoccupied guano island, anywhere, and incorporate it into the United States. To this day, there are still a few governments asking for their islands back. The Chinchas Islands were all Peru's, though, and by the mid-1860s the guano trade made up almost 60 percent of Peruvian government income.

More money, more problems, of course. Spain had never recognized Peru's 1821 independence and was looking to reassert some level of dominance. After some financial and diplomatic bullying that Peru did not submit to, Spain occupied the Chinchas Islands in 1864, recognizing their economic value. This was the opening act of the Chinchas Islands War of 1864–1866 in which Chile, Ecuador, Bolivia, and Peru allied against their former colonial master. Despite naval superiority, Spain had no invasion force to land and nowhere to resupply. The two-year conflict ended when Spain pulled its forces out of the Pacific back to Europe by way of the Philippines.

While the guano islands off the coast of South America were some fertilizer worth fighting over, the Chinchas Islands War was a relatively small engagement. The fighting over the nitrogen found in the Atacama Desert would be a different story.

The Atacama Desert is the driest desert on earth. Scientists think the region received no significant rainfall between 1570 and 1971 and that it has been extremely arid for around two hundred million years. This lack of rainfall created a region rich in sodium nitrate, known also as Chilean saltpeter. Potassium nitrate was found in abundance as well. Miners started exploiting these reserves in the 1820s, and the region would come to dominate the global fertilizer market for the next century.

Many Chilean miners worked in the saltpeter mines despite the fact that much of the Atacama Desert was located in Peruvian and Bolivian territory. The Chilean-owned Antofagasta Nitrate and Railway Company had an especially large stake in the region. Bolivia had agreed in 1874 not to increase taxes on Chilean interests for twenty-five years, but in 1878 the Bolivian government imposed higher taxes on Antofagasta Nitrate, retroactive to 1874. The company balked, and so Bolivia moved to seize its assets. In response the Chilean army crossed the border and occupied the Bolivian port city of Antofagasta in the southern Atacama Desert.

Needless to say, at this point tensions were running high. Fifteen days after the occupation of Antofagasta by Chilean forces, Bolivia declared war. Peru, bound by a secret alliance treaty with Bolivia, declared war on Chile little more than a month later. Thus began the War of the Pacific, a bloody conflict that would drag on for four years. Chile ultimately defeated its neighbors and gained large swaths of territory from both Peru and Bolivia. Before the conflict, Bolivia had access to the Pacific Ocean but would now be entirely landlocked.

Most historians have argued that the underlying cause of the war was the desire to control the vast nitrate deposits of the Atacama, plus some bonus guano islands. To this day

Bolivia pressures the Chilean government to allow it a sovereign access point to the Pacific Ocean, to little effect. The saltpeter mines gained in the conflict would be an important part of Chile's economy until the 1940s, when synthetic nitrogen production made mining the Atacama for sodium nitrate less profitable.

Literally tens of thousands of people died in these conflicts, and all this bloodshed was because people wanted to give their crops a little more nitrogen. It emphasizes just how important this plant nutrient was and still is for food production. With the advent of synthetic nitrogen, humans were able to break free of this natural constraint that limits productivity in many of the world's ecosystems: access to biologically available nitrogen. Scientists estimate that synthetic nitrogen fertilizer is directly responsible for the addition of three billion people to current global population. Whatever your views are on the Green Revolution and the chemical nitrogen fertilizer that is its cornerstone, it is hard to deny that much of the world's population currently depends on this system of agriculture for survival.

The entirety of natural systems on land produce around 100 teragrams (Tg) of nitrogen annually. Humans have recently surpassed this, synthetically producing about 110 Tg of nitrogen each year for fertilizer. Certainly, synthetic nitrogen fertilizer is detrimental to soil health. It is responsible for enormous amounts of pollution, and it is part of a reductionist view of soil fertility. However, we can't ignore that we are currently utterly dependent on it. The oft quoted statistic is that half of the nitrogen in an average human comes from synthetic fertilizer. The process of synthetically fixing nitrogen requires a massive amount of energy, up to 2 percent of the global energy supply. If lack of fossil fuels were ever to halt nitrogen fertilizer production, agricultural production would plummet catastrophically.

Have we reached such a high level of fertilizer consumption that there is no alternative to synthetic production that could ever provide enough nitrogen? The Atacama Desert, which at one point was responsible for two-thirds of global fertilizer production, produced a laughably small amount compared to what we now need. Organic fertilizers such as fish meal or kelp meal make up an extremely small fraction of current fertilizer use, and we would quickly destroy these resources if we ever tried to pull the millions of tons of nutrients we need out of them. If we are ever going to come up with alternatives to synthetic nitrogen, it will require us to scrape together as many sources of fertility as we can and dramatically alter farming practices. Recycling human waste could provide roughly 20 Tg N per year and composting programs across the globe could provide another 15 Tg N. Increased use of legume cover crops will help. Currently cover cropping produces around 40-70 Tg N, but if it were practiced on every piece of arable land (admittedly a tall order), a whopping 500 Tg N could be obtained. All soil nutrients, including nitrogen, should be viewed as precious resources that we cannot let slip through our fingers. This means comprehensive biosolids, composting, and other nutrient capture programs that recycle nutrients back to farm soils. It won't be easy to meet our nitrogen needs without synthetic production, but now is the time to start figuring this out.

EARTH ETUDE FOR ELUL:
Ready for Withering Flowers

SARAH CHANDLER

I'm familiar with your story
This gratitude you cultivate helps ground you
And yet, do you really deserve to ask for more?
The answer to this question will give you the balance you seek
Sometimes you need a reminder that we already said farewell to the month of Av
As it is written in Job: "Man born of woman is short of days, and fed with trouble.
He blossoms like a flower and withers, and vanishes, like a shadow."[1]
In Elul, you are instructed to enjoy the ephemeral beauty of the flowers without
worry of their withering
Since t'shuva/repentance is the name of the game, instead of fearing change we
welcome it in
Every morning the shofar calls you to t'shuva
Are you listening?
How might you be more awake in order to hear its sound?
Allow the August blossoms a chance to bring you to the presence you desire

Step 1. Gather flower petals into a large bowl—ideally four colors and four different species. The bowl is ideally wood but can also be glass or metal. In New England this is a great time of year to find a diversity of lilies, Queen Anne's lace, chicory, and aster.

Step 2. Fill your bowl with water covering the petals—ideally springwater, but tap water is also fine. The chance to visit a river, lake, or small spring will only add to the ritual.

Step 3. *Ask for something*. This is for real. If you're going to open up enough to do real *t'shuva* this year, you have to acknowledge that you are not yet whole, that there is something about yourself you want to change, or at least cultivate. A useful formula is "May I be …" or "Let me be …"

Step 4. Pour the entire bowl of petals and water over your head and proclaim: "*Horeini Ya Darkecha*—Reveal to me your path."[2] This is both the sealing of our request and a letting go of wanting only one thing.

NOTES

1.   Job 14: 1–2.
2.   Psalm 27:11.

SUPPLY'S DEMAND

GRAISON S. GILL

It is my hope, as a food producer in New Orleans, to cast some light on the shadows at our dining tables. Our supermarket shelves may be full, but our landscape is empty. Louisiana is a national leader in obesity, diabetes, heart disease, incarceration, violent crime, poverty— is that relative to the dwindling farms we have? I know so. Gandhi said that you could judge a nation by the way it treats its animals. And we can judge Louisiana by the way it lets one in six adults remain hungry. According to the U.S. Department of Agriculture, the top three crops in Louisiana are soybeans, cotton, and feed corn. None of those are edible in a state where one in four children are hungry.

Food has gotten farther and farther away from our stomachs, from our mouths, from our communities. We are no longer a city that celebrates its foodways, but its tastemakers. Do we praise Wonka or his candy? I bow to the sugar that made it all so sweet.

But we are now in a time beyond blame. There has been massive, shameful failure that has led to this current womb of crisis—but maturity teaches above finger-pointing. New Orleans needs cures, not more diagnosis. We must embrace truth before wisdom, now. We must begin renovating our soil and food system before renovating more restaurants and markets. Demand for food will always be, but supply is a delicate, delicate web. Our tax money does not need to subsidize consumption, but should encourage production. The plate is a mirror we look into three times a day. And if we do not begin to value the content of that meal, rather than its glamour, then we will paddle soul-deep into Narcissus's pool. No soil, no farms; no farms, no food; no food, no restaurants. Soybeans and cotton and feed corn won't taste good in even the best kitchen.

Successful social movements do not wait prostrate for the epiphany moment of hope and change. Freedom's angel is agency: the ability to do and not to do what one sees fit. The civil rights movement, the women's liberation movement, the LGBT movement, and decolonization were all born when policy, both public and private, failed. There existed a vacuum where we as a community failed our neighbors, and these cathedrals of justice were born in that failure. Assertive in their demands for equality and inclusive access, they provided what the government could not or would not. The food rights movement is even more potent, more powerful, more democratic: everyone in the world puts food into their body, every day. Fresh, healthy food is a human right, not a commodified privilege. Producers and eaters must seek to realign the paradigm and draw it away from consumption; we must begin again to celebrate production, to encourage the harvest, not fetishize the consumption and the fashion of consumers. Local and regional food is sincerely better for everyone and everything. Think about it: we do not import our music, our culture, or Mardi Gras. So why do we import our food, that most sacred altar in New Orleans?

CONTACT DIFFUSION:

Why Whole Grains Matter

GRAISON S. GILL

You never change things by fighting the existing reality. To change something, build a new model that makes the existing model obsolete.
—Buckminster Fuller

R. de Salis

Whole grains—regionally grown, stored, stone-milled, and finished—are imperative to our health. To our literal health, our children, our ecosystem, our economy, our education, our businesses, our wetlands and water table, our jobs, and the integrity of our future. Complex carbohydrates—whole grains—provide humankind's most important foodstuff. Civilization was civilized by regional whole grains in Mesopotamia; it is no coincidence or irony that gluten is society's cohesion. The health of whole-grain diversity reflects our own well-being. Whole-grains contain high-quality protein, minerals, omega-3 fatty acids, monounsaturated fat, dietary fiber, polyunsaturated fat, vitamins E and C, zinc, phosphorous, folic acid, magnesium, B vitamins, iron, and potassium, among other incredibly important nutrients.

Whole grains are the building block of the dietary-nutritional and financial economy of their region. Like carbon, whole grains are the bedrock of life: every living being has access and ability to fix carbon, but not every human being has access to whole grains. For these reasons, Bellegarde is continuing the work of reestablishing a regional grain economy. The French immediately established a regional grain economy when colonizing greater Louisiana. Illinois Country (Haute-Louisiane) encompassed present-day Missouri, Illinois, and Indiana. This fertile land, sparsely populated by Europeans, was an Eden for grain (corn, wheat) production. Grain grown, milled, and then barged down the Mississippi to New Orleans set the infrastructure precedent still in place today. Nearly sixty billion pounds of grain, two-thirds of America's domestic crop for export, still travels that muddy highway.

Bellegarde is investing in a very large American-made stone mill that will arrive this summer; its size will drastically increase our milling quantity and quality. Whereas we currently only mill enough flour for our bakery, we will soon be able to provide flour for area restaurants, bakeries, and the public. Most importantly, we are working to build an understanding between policy and people; we want our neighbors to know the importance of and the relationship between health and access to fresh food and politics: the democratization of fresh whole-grain bread—financially and physically—is the core principle at Bellegarde. Every day we strive to educate consumers about the benefits of whole grains. But as bakers, the only viable argument we have is taste: stone-milled, whole-grain bread tastes incredible. Thankfully no language can argue with flavor.

Bellegarde recognizes the following guiding values on whole-grains:

1. Health: Stone-milled whole grains provide essential nutrients and vitamins to the human body. Commercially grown and milled flours, processed into French bread, pasta, white rice, and cookies—pure starch—are culpable for Louisiana's staggering rates of obesity, diabetes, and heart disease.

2. Ecosystem: Irrigated grain bred for flavor and performance—not yields, not animal feed—is best for our land. New Orleans is the spleen of America; culturally, we posses the best elements from the river. But, Louisiana receives all the toxic runoff of America; like a barium test, corporate poison

slinks through the capillaries (the Illinois River) and veins (the Arkansas and Ohio rivers) and arteries (the Missouri River) of our Mississippi. Dredging for oil has ruined our wetlands, and dredging for high yields has ruined our soils. Regional, organic grain systems require less toxic inputs and produce better results for everybody: the farmer, the miller, the banker, the baker, and the folks downstream.

3. Artisans: A regional grain economy will encourage the reestablishment of craftsmanship in the disciplines of milling, baking, retailing, and their attendant professions. Clean, fresh whole grains will create numerous institutional, financial, and cultural roles as the relationship to our food is restructured and resurfaced—lost trades, disciplines, vocations, and traditions will enjoy a renaissance. Permaculture and stewardship will be invigorated as empiricism and empowerment are brought back to life because of organic and traditional farming methods. We seek to put the culture back in agriculture.

4. Energy, security: We eat oil as often as we eat food. Regional grain will sustain a transfer from fossil foods to real calories. Without artificial price supports (federal "farm" subsidies), the real cost of food will shift the demand away from ethanol. Producers always receive premium prices for their products when it is bought in direct markets. This will stabilize the stomach of the artificial commodities market. Remember the rising cost of bread when wheat made a run on the market in 2008? Despite record harvests, the financial markets artificially inflated the price of grain due to speculation, cornering, and other nebulous massage points. I eat bread, not five-year bond notes.

5. Supply and demand: Food consumption in New Orleans rests on lopsided shoulders, and it is giving Atlas scoliosis. Catering only to demand isolates and stifles supply. You can see that reflected in the dearth of famous products with local flavor and cultural terroir. Tabasco makes sauce with peppers from Central America; Cajun Boudin is made with pork from Canada; restaurant grits are made with corn from Iowa; Louisiana cane sugar is bleached and packaged in Yonkers, New York; gumbo is served with Arkansas or Texas rice; Chinese crawfish are in too many freezers. Regional organic grain will cement a relationship between producers and consumers because it draws the knots closer to the laces. It will shift the paradigm from tenuous, distant origins to a traceable and transparent source. By compressing the physical and emotional distance in which our food is grown, a relationship that fosters the fibers between suppliers and demanders will engender a healthier environment for all.

6. Integrity and ingredients = flavor: Fresh food, made with fresh ingredients, tastes better. Grains selected for their flavor and nutrition, as well as their milling and baking qualities, provide the precedent for such considerations.

AMARANTHUS LEUCOCARPUS *S. Wats.*

THE HIDDEN HERD

GRANT RICHARD JONES

Mosquito Creek, North Okanogan Valley, Okanogan River Sub-Basin

The creek was low but cold as a buried ax.
Fifty tiny steelhead twitched the bottom below the dam,
Size of my little finger, three inches long.
"Let's wade the creek back to the waterfall today.
We can look for the babies of the three big steelhead we carried
Over the spillway to our pond this spring," Chong said, smiling.

Stopped at the barn. Checked the freezer from the 2nd Hand—
Seven degrees, just right for the Sockeye we salted
In Hanna and David's root cellar in Havillah, now fermenting.
At the pond sat down, pulled on rubber boots,
While Chong, lifted out the thermometer, a cold 53 degrees.

Dogwood and willow wove in our arms, scrapped
Our ears, tangled our legs as we hopped the bars.
Each shallow pool held no wiggling fry,
But high sand cliffs below the granite waterfall
Were gouged by a summer of skidding hooves, tumbling.

So we climbed on our knees, scrambling high to the trail.
A boulder held our weight as we rested, catching our breath.
Then we heard the stones rolling, splashing into the creek just
Like yesterday's earthquake had shuddered our breakfast.
Was Mt. Baldy quivering, again, fourteen miles north of us?

Then we saw them coming, around the waterfall, scrambling,
Hopping and sliding the sandcliff to drink the cool water,
The whole lost herd from Wild Horse Springs on Mt. Hull;
One after the other they squeezed in down below us so calm.
Looking for steelhead we'd found fifty ungulates on vacation:
Bighorn sheep, *Ovis canadensis canadensis*.

FALLING IN LOVE WITH YOUR PLACE:

Caring for the Land and Each Other

GRANT RICHARD JONES

Advance confidently in the direction of your dreams…
Endeavor to live the life you have imagined…
As you simplify your life,
The laws of the universe will appear less complex…
—Henry David Thoreau, *Walden*

Where I live is the heart of back-to-the-land organic orchards and open-range ranchlands. Several small town necklaces follow the streamways: Brewster, Bridgeport, Nespelem, Inchelium, and Kettle Falls follow the Columbia; Keller, Aeneas, and Republic up the Sanpoil; Monse, Malott, Okanogan, Omak, Riverside, Tonasket, Oroville, and Loomis up the Okanogan-Similkameen; Pateros, Carlton, Twisp, Winthrop, and Mazama up the Methow; and Chesaw, Malo, and Danville up the Myers and Curlew of the Kettle. These are the supply valleys for organic fruit, vegetables, and meat to the Puget Lowlands, the Salish Sea megalopolis. It's also one of the most culturally diversified belief landscapes in the West. Whereas the Salish Sea urban metropolis is uniformly democratic, North Central Washington is American Indian, Mexican American, Buddhist, Christian, agnostic, hippie, escapist, liberal, conservative, cowboy, logger, rancher, orchardist, organic farmer, artist-singer, writer, teacher, doctor, rich, pension-retired, and dirt poor, all living in harmony with each other in positive tolerance of each other's values. Wherever you are, I hope you've become a river, your own river, because there's nothing like you.

Landscapes have always reached out to me. I feel the passion for life inside them, and this has given me the words to become very attached to the landscapes of my Mother Earth. Like all living things, landscapes need partners. They need caretakers to appreciate their expressiveness and stewards to reciprocate with them to increase their energy and fullness at every scale, from garden to region. I put this book together to celebrate my love affair with the landscape and to capture the magnetism possible between its partners.

Simple things become most sacred. The overhanging tree becomes a haven for your deepest feelings. A valley where you live can arch your heart and soul and even transport you wide awake in your dreams and in that way your valley will become a link to distant places where you can be with friends. You learn to talk with other living things like trees that are always waiting and how to interpret their stories. You can talk with those who are gone and honor their spirits still residing under the trees in certain places. You've become a bridge and never have to leave to go anywhere.

With age, our muscles can weaken and become laced with fat, or they can get more sinewy through each landscape experience, braided with cords that got stronger then relaxed in their length from all the turns you made. Also, as the years go by, you find you can pull off to rest and sleep anywhere, wake restored, and expand to breach old barriers and discover new channels. Your male and female halves, also entwine like rivers and clouds do, feed each other and restore each other. Every piece of every landscape is part of a watershed, every surface faces downstream

out to the sea (some inward to dead seas). Gravity swings everything, including you, makes your clock keep ticking. You're also a tributary in your own community watershed. I moved out of the city because it was becoming generic and it's new people more intolerant and regionally dumb, looking to imitate, not be its unique self. It was starting to eat itself like a carnivorous flower. I felt like I was dying. I chose to settle into a more diversified culture in a physiographic region with colliding ecologies. It's where the largest county in Washington, Okanogan, joins one of the smallest, Ferry. It's the Okanogan Country, where the North Cascades Subcontinent collides with the Okanogan Highlands Subcontinent and crushes it against the Northern Rocky Mountains, the original Pacific shore. All this happens within a hundred miles. Coniferous rain forests, alpine meadows, aspen springs, tamarack parklands, ponderosa savannas, sagebrush steppes, bunchgrass prairies, bitterbrush deserts, cottonwood galleries and river grasslands are tightly woven in this rumpled network of ridges between sequestered hidden valleys.

Each landscape offers its own story. Each landscape is a book that's open to any observer who's awake. Each landscape carries the language to describe itself; in fact, the biologic foundations present in the landscape are the vocabulary of the local language in our own place. Everything you can say about your place derives there and has its origin in what you can see.

The rocks and the trees, and the rivers and animals in the landscape, are the words and phrases, sentences, and paragraphs. Together these rocks and words make up the full discourse of your exchange with it. If you don't know its features, you can't converse with it. In other words, you're a partner at the table with your local landscape, and if you're open, you can hear each other. The more you see and describe, the better you can listen. The landscape's spirits will weave with the life energies inside each one of us if we are awake with her in the present moment.

It's for these reasons that I see and hear everything around me as a poetic structure. I see the landscape of a place as the architecture of a poem. Each landscape makes the language, and the language it makes can save it. Each landscape depends on you for its survival, and your life depends on this relationship. You're its steward and interpreter. To communicate our feelings toward those whom we love, the landscape we share gives us the full range of metaphor and depth of meaning we wish to celebrate in our human relationships.

Chong Jones

A CEREMONY OF NECESSITY:

Field Notes from an Urban Farmer

CHARLOTTE X. C. SULLIVAN

Standing in the same place for eight hours in Union Square is a little like going to the beach all day. Only, instead of watching waves of water, you watch waves of people. Fridays I arrive around 7 a.m., set up, and begin to unload the usual supplies from the back of the pickup. The tables, the chalkboard, its neatly listed offerings and prices. I keep the typography and palette of our signage understated so that it doesn't interfere with the beauty of our food. Wooden crates and brown baskets are taken out of the truck's cab, along with burlap to cover the tables, and colorful tablecloths to cover the burlap. Lastly, out comes the scale, my apron, small bills, and of course, whatever food we have brought to sell. (I often joke that farming is "nothing more than moving piles around from place to place.")

The produce we sell depends entirely on the time of year. These past few weeks we have brought pounds and pounds of greens. Planted in cold frames last fall, they are finally, after the long winter—a winter that everyone still seems to be grumbling about—ready to be picked: spinach; two kinds of mustard greens, tatsoi and golden frills; and some baby red Russian kale. Soon lettuce will arrive, along with nettles, and maybe small amounts of sorrel, beet greens, lovage, and foraged chickweed. These greens are almost entirely managed by hand, from the time the seeds are planted to the moment they are transplanted into the ground or cold frames, harvested with serrated, red-handled knives, and finally washed and spun dry in an old washing machine. Not only has the produce been picked within twenty-four hours of being sold, it has also likely been washed, packed, and propagated by me.

Queens Farm and Union Square are only seventeen miles apart, and yet the landscape of these two places couldn't be more different. The farm manages to retain a sense of expansiveness and abundance, despite being surrounded by suburban sprawl. The market allows me the opportunity to communicate these characteristics through my display of produce and exchanges with the public. I'll position items near one another to inspire new recipes. The colors of the tablecloths are chosen to mimic the time of year, or the weather forecast for that day, or even the tonal range of the food we're selling. One of my favorite tablecloths is the exact pink of a new rose, but paler. It is a color that seems to suggest the almost-in-full-bloom atmosphere of April. It is the pink of new love, not old.

I believe food is not only meant to be consumed, but contemplated and considered. It is one of the simplest ways we have to reflect on the life cycles—of our own lives, and the lives of those around us. If we choose, food can be much more than sustenance for survival—its necessity makes it the perfect basis for ceremony.

OCTOBER

LAND AND OWNERSHIP

PREDATORS

ABBY SADAUCKAS

When I brought home my first chickens, it was as if I put up a sign that read, "Welcome Predators!" My homestead, tucked into the woods, was a haven for all kinds of wildlife. I acquired the hens from the farm where I worked. They had ordered a truckload of pullets, and the three that came home I named Quiche, Omelette, and Scramble. I housed them into a hastily constructed A-frame chicken tractor. Within the week Scramble was found, still inside the enclosure, with her neck wrung, the culprit thought to be a raccoon.

Several years later I was renting a farm with a large renovated farmhouse sitting atop a fieldstone basement. The chicken house, sited just down the hill, backed up to the woodpile, and one evening in late fall I saw a half-brown, half-white weasel dart out of the pile and into the barn. I didn't think too much of it until a week later, when I opened the hens in the morning to find three of them headless and bloodied, lying on the floor of the coop. The weasel was the obvious culprit, and so ensued a six-week battle as the killer whittled down my flock. I awoke one night to find that the weasel had nearly decapitated the rooster, General Tsao. I laid his body down in the snow, hotly pursued the weasel, and returned later to find the general was still alive. I place the general and my favorite remaining hen in the farm's ICU, a cardboard box in the corner of the kitchen. Exhausted from several nighttime forays out to chase the little devil, I got back into bed. Within minutes I heard the sounds of a chicken in distress and dashed down to the kitchen. I arrived in time to see the weasel climbing into the ICU box with the general and my hen. Dashing in on my heels, Chicory, my terrier, took after the weasel. Standing in the doorway I realized the weasel was coming straight for me, and so I jumped up onto the counter in my bare feet. From this vantage point I followed the progress of the chase into the hallway, where the weasel, confronted by the stairs, let loose his scent glands. Chicory answered this affront by grabbing the weasel in his jaws and shaking him. As he did so, the life left the weasel, and I jumped down and threw open the door for Chicory to trot through, depositing his prize outside.

Dogs aren't the only effective means of predator control, in my experience. This past season a local owl decided our farm was a buffet, serving up meals between midnight and 5 a.m. Our broilers, raised on pasture in open-sided A-frame structures, became the owl's primary target. Each morning there would be one dead broiler, pulled out of the house a distance away from the others, its head and breast meat having been removed. First, additional fiberglass posts were put up to deter the bird from flying in, but this did little. By day three or four of this I decided to put our geese on sentry duty. At dusk I drove Hank, our gander, and his two geese into the broiler pen, putting out a bit more grain to keep them entertained. The next morning the broilers were unharmed. The geese continued as chicken guards until the broilers grew large enough to be safe from harm. However, when later in the season our turkeys drew the notice of the owl, we moved the geese to the turkey pasture, and again the attacks ceased.

Geese are often thought to be loud and mean to farm visitors, but on our farm they have come to be indispensable for keeping avian predators out of our pastures. As for Chicory, though his exploits in the compost pile make him unpopular, and the UPS driver wouldn't hesitate to run him over, he has cemented his place on the farm with his bravery in the face of that tiny yet terrifying weasel.

A YELLOW HOUSE IN AMES

ERICA ROMKEMA

Lemons, Bosc pears, butternut squash. Oranges and carrots. I don't notice until I get to the register, sometimes not until I get home and pile up my produce on the counter. Then the colors find each other, a spectrum of shades. Not long ago a friend came to visit, and I gathered red: red onions, red potatoes, red D'Anjou pears. We had pears and cheese and crackers as we cooked the red meat and red onions and red potatoes into a spicy-sweet Pakistani kima. The sky fell into twilight but we kept the kitchen warm.

We eat in this house, a yellow hundred-year-old Victorian

Flaming June

Frederic Lord Leighton

in the Ames Historic District. Five women, all exploring the stages of our lives, young still, searching still—more, or less than before? Tomoko can spend eight hours a day in the garden, despite the summer heat. Her study involves community and urban gardens as a solution to social and environmental problems. Sue brings the farm to Iowa State University— literally—by working with local farmers and the ISU dining staff to get food from Iowa, for goodness sake, onto the plates of Iowa students. Rachael, our undergraduate, comes back from Ecuador with tea and startlingly smooth, rich chocolate. She wakes early to bake scones, stays up late eating pepper and avocado slices with a friend. Emily spends days whipping lattes, nights swimming, weekends playing music—and in the midst tries new recipes: pesto, bruschetta, baked apples. She practices violin, the notes moving from urgent to soothing, as we stir soup in the kitchen. And me—I read my way through the MFA program, with pear crisp baking in the oven. I add a minor in sustainable agriculture so I can get into it all, the food and land and people, this bright and welcoming mud-on-our-hands network.

Agriculture leaves fingerprints on every facet of our existence. And at the end of a day we come home to how we eat. We eat in all the richness of a pesticide-free squash, butter sliding into the dips and crevices, fair-trade brown sugar glinting against that yellow-orange of flesh so bright it's like Frederic Lord Leighton's *Flaming June*. We eat the tartness of chard in late spring, bite four kinds of lettuce as we forget the names in the mix. We eat the depth of bread, Tomoko working the starter into yet another flavor.

Honeycomb from that place north of town. Butter from Picket Fence Creamery, to our west. Cherries from a tree in the back yard; we didn't plant it, but we know its bounty, some blessing from a former resident, we the happy receivers.

In summer, I stand in a skirt on a ladder and reach among the branches. Is this how Mary Oliver felt, her bear-self harvesting the richness of blackberries, in her poem "August": "thinking / of nothing, cramming / the black honey of summer into my mouth"?

I imagine the someday of children running around my feet, shouting as I toss down cherries. Maybe they will catch them in silver pails, like the girl in Blueberries for Sal, to collect before we return to the kitchen. My children will bite the fruit and red juice will stain lips, pits spit to the ground, so many laughing squinting eyes. Will their eyes be green, like mine and my mother's?

I can only wait and see. In the meantime there is so much good in imagining, in gathering, in being. In this house we share our lives around the table. Sometimes—it's true—we argue about the curly-haired young men who seem bent on disrupting our existence. But mostly we argue inquisitively, cooperatively, around how to make the world more beautiful, more just, more gracious. And we try out ways to do so. Sometimes all that means is making a pot of tea.

When I leave this house I will miss its coral walls and shared recipes and laundry line. I will think of merry dinners and icy sidewalks, the neighbors' goat on our roof, the sound of the university bus braking to a stop. Every day we cross the porch beneath Tibetan prayer flags, opening our arms to a vibrant world.

LAND FROM THE SEA

DANIEL TUCKER

I always wanted to live on the land. A nomadic upbringing instilled in me a strong homing instinct. Early years of homeschooling in the Oregon rain forest inoculated me with a hunger for fresh air and a distrust of institutions.

After high school in Homer, Alaska, I stumbled into commercial fishing as a summer job. I made $30,000 that first season, and I was hooked. It's a great moneymaker for back-to-the-landers: you get a lot of exercise, learn about everything from hydraulics to rigging to radar, how to eat a sea urchin and skin an octopus, how to work twenty-four hours a day, and how not to get killed by heavy, sharp, and fast-moving objects. It's 90 percent drudgery and 10 percent glorious adventure that you will never, ever forget. And the terrifying moods of the sea will make you crave the embrace of Mother Dirt like nothing else.

After an uninspiring fling with college and city life, I returned to Alaska to buy land and build my homestead. My first step was buying a twenty-foot-diameter yurt from my friends' yurt company in Homer, Nomad Shelter. A word about the yurt: it is cozy, stout, and has taken excellent care of me for six years now, in Alaska, in Cascadia, in the town, and in the country. I highly recommend a good yurt. With it you can act like an invasive species, colonizing disturbed landscapes overnight, and quickly moving on if necessary. Genghis Khan conquered the world from a yurt. Just as importantly, the circular space with a central skylight feels eternally wholesome, a wheel, basket, nest, and lens for the soul.

So I bought land. After my second season, when I made only $12,000, I found 9.68 acres of spruce and fireweed in the hills above Homer and got owner financing on it for $45,000 (land is cheap in Alaska). The loan was daunting, but I loved the land and felt it was my true purpose. I loved skiing out the front door of the yurt, following tracks in the snow, the stars, the fire, the silence.

The third season we hit it big. Commercial fishing is a casino, but as a deckhand your only investment is time, and if you stick it out long enough, you'll probably get in on a good season. Good prices plus lots of fish means I made $60,000, before taxes, and paid off the land. I had never felt better in my life.

However, I was possibly the only twenty-one-year-old living in Homer at the time, and another winter alone in the yurt was not appealing. I headed south to Cascadia, where my Homer High diaspora was playing in bands and pursuing other glamorous pastimes. The currents of life took me back to Bellingham, to a hippy-disciplinary liberal arts college, the only college I could stomach. I was there more to be around interesting people my age than anything else. I turned down student loans, against my grandparents' advice, and dropped out when I could no longer justify the money I was giving to a corrupt institution. But by then I had found my people: farmers, boatbuilders, clay diggers, artists.

Alaska faded into the distance. After years of enduring subarctic winters, I loved the mild climate of Cascadia. I discovered plum trees, fig trees, and year-round kale. These are big things for a homesteader. I brought the yurt down south to colonize an acre of land owned by a friend's grandmother, right in town. The subarctic homestead was left to the porcupines.

Daniel Tucker

For almost two years we lived in a blossoming ramshackle eco-village, until one day the letter arrived from city hall. The "illegal" yurt was exiled to the countryside, but Trausti's cob cottage remains. The experience taught me a great deal about cultivating both soil and community. My friend's family gardens there still. I was now determined to buy land in Cascadia, where I could plant walnuts and mulberries to last for generations. But land in this wonderful place is expensive.

There was endless talk of collective land-buying in hazy living rooms over bowls of quinoa with nutritional yeast. But no one had money. At one point I suggested we all get fishing jobs in Alaska, but people were too comfortable fishing pizza out of dumpsters and going on naked bike rides. It's a good life, and I don't blame them. I kept fishing.

In the end it was my father, a forester living in Ireland, who had the desire and the funds to help start a living laboratory for agroforestry. I found five acres of sunny hillside with a seasonal creek on San Juan Island for $100,000, and we split it fifty-fifty. Now there's a pond, a deer fence, and twenty-two chestnut trees, and I feel like I'm getting started. I must credit the sirens of the island for luring me out here. Years ago, crossing the Gulf of Alaska through hammering seas, I swore like Odysseus that if I survived, I would march straight inland and plant my proverbial oar in the dirt and sow seeds in the earth. Here I am.

FIG.VII.

Art .143.

Spherical Harmonic of the third order.

Fact: 8.5 million hectares of land in Europe is still managed as commons.

YOU MADE FIVE QUAIL

STEVE SPRINKEL

You made five quail this season,
down from seven nice birds last year.
I watched you two race around in the old kale,
audibly concerned about where you could lay.
Your little startled chip-chip-chips mirrored how critical
our mutual struggle looms against the tired dust.
That's why I didn't mow that block down,
having enough priorities crashing in from all sides.
Did you fly in here so dark and scrawny,
oddly matte, with a droopy, little hussar's plume?
Just for you, I neglected one hundred million weed seeds,
just for you we watered that chest high stand.
You are welcome to what we let go wild,
for we expected you to visit that very spot again.
Of course I will wait until all of them can fly
on their own, wherever you sail off to in this life.
I will wait for you for as long as I stand here,
even if you're late, I always will shape a shelter on your behalf.

THE NEW FARM FAMILY

HARRIET FASENFEST

For me, taking back land and labor means living collectively with the clear mission of using the savings in the time and money to put toward our homes—both in the city and with our family farmers. Of course, living collectively is not a new concept. Students, artists, advocates of alternative living, and recently immigrated families have been doing it for years. They all know that living collectively can save money. Still, what distinguishes the urban-rural farm family collective is its mission. It is about coming together in the spirit of the populist movement when farmers and industrial workers fought the fat cats of yore. Is anything really all that different? Yes, missions are elusive and hard to hold fast to, but no more elusive than the hope we will be able to thrive, or just survive, in the years ahead as the costs of living and growing our food keep barreling down on us. If we do not get creative in imaging lifestyles outside the demands and "logic" of the marketplace, we will all—urbanites and our farming friends—be caught on a treadmill of market mentality with its inevitable diminishing returns.

An excerpt from Harriet Fasenfest, *Remembering, Reenvisioning and Returning Home: A Curriculum Guide to the New Home Economy,* forthcoming.

LONG OVERDUE LETTER TO JEFFERSON, 2016

JOHN FRANCIS MCGILL

Sage of Monticello, part
farmer, part pen   of
modern revolutions,

I beseech your future visions,
please give sketches

                    to
our
surviving present, now engrave

these contrapasso winds—

Can you do that
   for me,
          an American,

          Good old
Jefferson?

 The dust upon
your fine little colt
          might be the grace
of the new American.

               I
remember
at Gettysburg He said,

May we stand on
         the same
              ground,
and as I write you now

(still Abe knows by heart
mother's closing prayers),

my dear friend, I think
to meet with you

up in the   Badlands or   out
on the Appalachians.

Let's beyond
            the Great Rock
Mountains

climbing over hurdles
gargantuan.

                    I am here in my
shoes.
I am here waiting to hear
again,

  when

Liberty rings out to sing once
more

that the earth belongs in
usufruct to
            the living

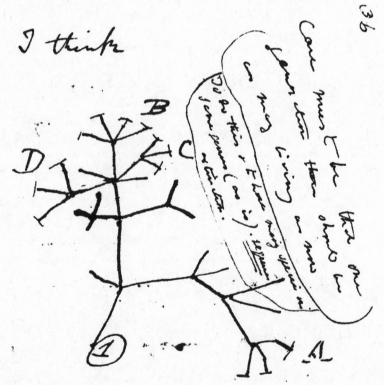

Darwin's First Tree

NOVEMBER

GENETICS

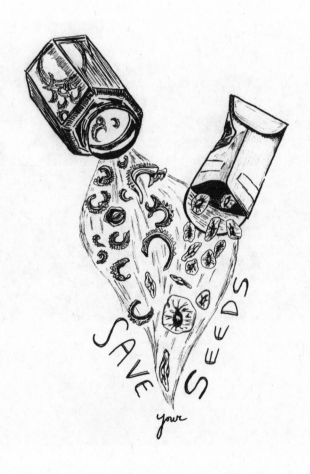

SAVE your SEEDS

WINTER

MARÍA JOSÉ GIMÉNEZ

come unhurried
without stumbles

come cascade of quiet
bark and autumn leaves

come glint of indigo
core of cloud

come ponderosa sweat
juniper breath come

come sleepless night
come run rumble come

the barren hills are deep in slumber
sweet with maple crowns come

darkness has gathered the day
the moon rests in frozen eyes of water
and the birches bow subdued at your feet

WHY TRADITIONAL BREEDS?

CHRISTINE HEINRICHS

Louis Ducos du Hauron

Traditional pure breeds carry important irreplaceable genes, the value of which remains for future events to determine. They are also beautiful, productive, will reproduce your flock, and connect you to history.

Back in August 1910, American poultry advocate C. F. Townsend made the case for "thoroughbred" poultry because of its broader economic value. After the eggs and meat are sold, the hatching eggs and stock are valuable to other breeders. The uniform shape and color of the eggs bring a premium price. "Big money can be made and is made in breeding good stock," he advised.

This economic incentive isn't the same as it was in those days, when meat and egg producers were also show exhibitors and advertised their prize-winning birds. Today, the irreplaceable genes are one of their important values. Traditional breeds may rescue the poultry industry of the future.

Breeds are the repository of genetic diversity in domestic livestock. Each pure breed, such as the English Dorking, with its unusual five toes, has a unique appearance, productivity, and behavior. They breed true, which means that when they are mated together, their offspring are predictably like them. Orpingtons always produce more Orpingtons.

Landraces, such as Sumatras, are local or national breeds that develop in a geographic area. They're influenced more by natural selection than intentional selective breeding by humans. They develop in response to environmental pressures and conditions. They existed before people identified, named, and described them. Other old foundation breeds, such as Cochins, are the result of domestication and selective breeding going back centuries. They are the breeds from which composite breeds such as the Wyandotte were developed. Some of those have long histories, and some are modern. Breeders continue to cross breeds to develop new production birds today.

Breeds are a package deal, not a collection of individual traits such as comb type and body conformation. We cannot know all the traits that make up a breed. To lose a breed is to lose the entire unique genetic package. All chickens are the same species and share some genes, but other genes are unique to the breed. Ducks, geese, turkeys, and guineas similarly share traits within their species but carry others that make them quite different, both from other domestic breeds and from wild relatives.

Traditional breeds are part of a culture that is being fragmented and lost. Traditional breeds do not flourish in industrial settings. The traits that make them special include being good foragers, good brooders and good mothers (and fathers), alert protectors, longevity, disease and parasite resistance, ability to mate naturally, and fertility.

Traditional breeds are an important part of an integrated and sustainable farm. Each pure breed's characteristics suit it to a climate and certain production goals. The Chantecler, developed in Canada, is suited to a cold climate. Mediterranean breeds such as the Leghorn, the Ancona, and the Spanish group are known for egg-laying.

Sustainable, integrated systems include poultry as working contributors to farm ecology and production. They consume weed seeds and insects, the traits the chicken tractor uses to advantage. They consume green waste and produce high-nitrogen manure for fertilizer. They provide meat and eggs. They reproduce themselves and perpetuate the flock.

Because domestic breeds result from selective breeding, they are likely to lose their utility characteristics and value if kept purely as exhibition, hobby, or ornamental birds. The selective pressure of utility production is a significant influence on the genetic direction of the breed. The birds a breeder selects for the breeding pen reflect the goals of the breeding program. When that means egg or meat production, those traits will remain strong in the flock.

For the breeder, choosing which birds to breed is never simple. Flocks need variability to be vigorous and avoid the pitfalls of inbreeding. On the other hand, birds need uniformity and predictability to retain breed identity. Industrial strains seek uniformity. Traditional pure breeds seek genetic diversity within phenotypic consistency. Breeders spend hours observing their flocks, as well as keeping records of growth, egg production, age at maturity, and health to guide them in choosing which birds to put in the breeding pen.

In the twenty-first century, industrial chickens are controlled by a few multinational corporations dedicated to increasing profits from a narrow genetic base. While that succeeds in the marketplace, it is inevitably subject to failure. Such genetically similar birds are all vulnerable to the same diseases. The crowded conditions in which they are kept, made possible by subclinical doses of antibiotics in their feed, create conditions under which disease outbreaks spread rapidly and are often resistant to treatment.

Breed standards are mainly physical but also behavioral. Selective breeding is guided by breed standards. Conformation, plumage, comb, and color are all significant aspects of the description. Traits such as fertility, parasite and disease resistance, and longevity are less easily observed than physical traits.

Breed health depends on maintaining a viable population size in geographically separate flocks. Birds raised in different environments under the supervision of breeders pursuing different breeding strategies will insure a healthy, strong breed.

Hobby breeding can save rare breeds from extinction, but finding or creating a market for traditional pure-breed poultry will generate market conditions that give them a more secure future. If breeders can sell their birds and earn income, they will raise more of them. Having an

Katsushika Hokusai

economic purpose fulfills one of the original purposes of domestic poultry.

Traditional breed poultry need to be more than living exhibits in museums. Offering the public the option of purchasing traditional breed meat and eggs will assure the future of traditional breeds as well as good food.

Christine Heinrichs is the author of How to Raise Chickens and How to Raise Poultry, *which focus on raising traditional breeds in small flocks. Her latest book is* The Backyard Field Guide to Chickens.

Reprinted from Christine Heinrichs, "Why Traditional Poultry Breeds?," The Poultry Site, June 2, 2015, www.thepoultrysite.com/articles/3497/why-traditional-poultry-breeds.

HERITAGE CHICKEN

CHRISTINE HEINRICHS

HERITAGE CHICKEN

Frank Reese of Good Shepherd Poultry Ranch in Kansas, in cooperation with the Livestock Conservancy and others, has developed a Heritage Chicken breed definition. It's specific to chickens but is appropriate for all poultry products. His goal is to clarify the meaning for consumers. Other breed conservation and humane organizations have approved it.

1. American Poultry Association (APA) standard breed: Heritage Chicken must be from parent and grandparent stock of breeds recognized by the American Poultry Association prior to the mid-twentieth century, whose genetic line can be traced back multiple generations, and with traits that meet the APA Standard of Perfection guidelines for the breed. Heritage eggs must be laid by an APA Standard breed.

2. Natural mating: Heritage Chickens must be reproduced and genetically maintained through natural mating. Chickens marketed as "heritage" must be the result of naturally mating pairs of both grandparent and parent stock.

3. Long productive outdoor lifespan: Heritage Chicken must have the genetic ability to live a long, vigorous life and thrive in the rigors of pasture-based outdoor production systems. Breeding hens should be productive for five to seven years, and roosters for three to five years.

4. Slow growth rate: Heritage Chicken must have a moderate to slow rate of growth, reaching appropriate market weight for the breed in no less than fourteen weeks. This gives the chicken time to develop strong skeletal structure and healthy organs prior to building muscle mass.

5. Chickens marketed as "Heritage" must include the variety and breed name on the label. Terms like "heirloom," "antique," "old-fashioned," "old timey," "traditional," and "historic" imply "Heritage" and are understood to be synonymous with this definition.

ABBREVIATED DEFINITION

A Heritage Egg can only be produced by an APA standard breed. A Heritage Chicken is hatched from a heritage egg sired by an APA standard breed established prior to the mid-twentieth century, is slow growing and naturally mated with a long productive life.

Definitions by The Livestock Conservancy and reprinted from, "Definition of Heritage Chicken," The Livestock Conservancy, http://livestockconservancy.org/index.php/heritage/internal/heritage-chicken.

Julius Wiesner

Frustules of Diatoms

ORGANIC FARMERS ARE NOT ANTI-SCIENCE BUT GENETIC ENGINEERS OFTEN ARE

ELIZABETH HENDERSON

At one of the public brainstorming sessions for the New York Organic Action Plan, an organic farmer made an impassioned plea for support for "independent science" and told us that with 8.5 billion mouths to feed by 2050, we will need genetic engineering to prevent starvation.

I would like to examine these words carefully to decipher what they mean, how those words are used by this farmer and by others, and suggest how the movement for locally grown organic food in this country should respond.

What is the meaning of "independent science"? As co-chair of the Policy Committee for the Northeast Organic Farming Association of New York (NOFA-NY), I have been an active participant in the coalition that is campaigning to pass GMO labeling legislation in New York State. In this capacity, I have spoken at public meetings, to the press, and on radio interviews. A question that I have heard from proponents of biotechnology is, "Why do you organic farmers oppose science, like the climate deniers?"[1]

The first time I heard this, I was startled and felt defensive. Had I ever opposed science? I searched back through things I had written and reviewed all the policy resolutions the members of NOFA-NY had passed over the years. I found a few places where I criticized reductionist science and defended "indigenous knowledge," things like composting and crop rotations that people who practice a craft know and pass on to their children, and that has not been proven by research at a university. But nowhere could I find any statement opposing science. Just recently, I reviewed with approval this statement from an organic farming group:

We support the International Federation of Organic Agricultural Movement (IFOAM) definition that organic agriculture is a

Glyphosate degradation

production system that sustains the health of soils, ecosystems, and people. It relies on ecological processes, biodiversity, and cycles adapted to local conditions, rather than the use of inputs with adverse effects. Organic agriculture combines tradition, innovation and science to benefit the shared environment and promote fair relationships and a good quality of life for all involved.

My farm has cooperated in any number of research projects with Cornell University scientists. We have tested cover crops, held a field day with the Cornell soil health group to allow them to demonstrate the ways a farm can test for biological activity and use a penetrometer and a rain simulator that shows how much aggregate stability the soils have. We spent seven years

working with Molly Jahns and her team on breeding a variety of sweet pepper. It is earlier ripening, cucumber mosaic virus–resistant, and open-pollinated, and now bears the name of our farm –Peacework. I served for four years on the Sustainable Agriculture Research and Education Program technical committee and three years on the administrative council. Unlikely activities for someone who is against science.

So what does this question about independent science really mean? I have come to understand that by "science" the biotech folks mean genetic engineering. They are deliberately conflating these two terms. And that seems to be how the farmer at our meeting was using the words too.

So, since I do not oppose science, do I oppose genetic engineering? Yes and no. I share with geneticists their fascination with the functioning of

305

the tiniest of particles that make up living matter. One of my favorite books is A Feeling for the Organism,[2] a biography of Barbara McClintock (1902–1992), a cytogeneticist who specialized in corn. McClintock was one of the first to map the corn genome. She demonstrated that genes turn physical characteristics on and off and discovered genetic transposition or "jumping genes." She shook the notion that science held as a truth that the genome is a stationary entity with the genes in an order that is unchanging by showing that it is subject to alteration and rearrangement. For many years, the mainstream of science regarded her with disapproval, only eventually to catch up with her and then heap honors on her great discoveries. Science lurches forward—and a great leap is yet to be made for a full comprehension of the relationship between genes and the environment.

The more geneticists look into it, the more complex the relationship of genes to physical traits turns out to be. As Jonathan Latham puts it:

A defined, discrete, or simple pathway from gene to trait probably never exists. Most gene function is mediated murkily through highly complex biochemical and other networks that depend on many conditional factors, such as the presence of other genes and their variants, on the environment, on the age of the organism, on chance, and so forth. Geneticists and molecular biologists, however, since the time of Gregor Mendel, have striven to find or create artificial experimental systems in which environmental or any other sources of variation are minimised so as not to distract from the more "important" business of genetic discovery.

But by discarding organisms or traits that do not follow their expectations, geneticists and molecular biologists have built themselves a circular argument in favor of a naive deterministic account of gene function. Their paradigm habitually downplays the enormous complexities by which information passes (in both directions) between organisms and their genomes. It has created an immense and mostly unexamined bias in the default public understanding of genes and DNA.[3]

McClintock's story reveals how hard it is for the scientific mainstream to accept new concepts. This becomes especially difficult when large commercial entities like chemical and seed corporations build their empires on an interpretation of a scientific phenomenon. And even more difficult when our universities are starved for public research funds and become dependent on corporate support.

I am not against genetic engineering in principle, nor is the organic movement internationally. What we are against is the rush to commercialize crops that have not been adequately tested for safety. There is so much we do not know about them. When you move one gene, many other genes shift, and geneticists do not know enough yet to predict the results.[4] That is why standards for organic certification in the United States and all around the world do not allow the use of genetically engineered (GE) seed or other materials like GE rennet in cheese. In regulating any novel technology, we should follow the precautionary principle. Test carefully and at length before commercializing. This has not been done with GE crops. Every cultivar is different, and each one should be tested individually.[5] Meanwhile, corporations like Monsanto, Dow, and Syngenta have been commercializing a very few moneymaking GE cultivars that farmers are growing on millions of acres that are doused with toxic chemicals.

The scientific evidence shows that the widespread adoption of genetically engineered crops in the U.S. has led to: (1) an increase in pesticides used in agriculture, according to the U.S. Department of Agriculture's Pesticide Data Program; (2) development of herbicide resistance in over twenty weed species; (3) insecticide resistance in target pests, including corn rootworms; (4) increased residues of pesticides in foods, including Roundup, a probable human carcinogen; (5) loss of biological diversity, including Monarch butterflies; and (6) massive increases in seed costs for farmers.[6]

The EPA provides annual average use estimates for the decade 2004 to 2013. According to Carey Gillam:

Seventy crops are on the EPA list, ranging alphabetically from alfalfa and almonds to watermelons and wheat. Glyphosate used on soybean fields, on an annual basis, is pegged at 101.2 million pounds; with corn-related use at 63.5 million pounds. Both those crops are genetically engineered so they can be sprayed directly with glyphosate as farmers treat fields for weeds. Cotton and canola, also genetically

	1993	1999	2012	2015
Soybeans				
Grain	20	20	20	40
Hay	15	200	200	100
Forage	15	100	100	100
Maize				
Corn grain	0.1	0.1	5	5
Corn stover	NT	NT	6	100
Sweetcorn	0.2	0.2	3.5	3.5
Oats				
Grain	0.1	0.1	0.1	30
Wheat				
Grain	0.1	5	5	30
Straw	0.1	85	85	100
Edible beans	0.2	0.2	5	5
Alfalfa				
Dry hay	0.2	200	200	400
Silage	0.2	75	75	400

Figure 1. Glyphosate tolerance levels since 1993 (EPA)[11]

engineered to be glyphosate tolerant, also have high use numbers. But notable glyphosate use is also seen with oranges (3.2 million pounds), sorghum (3 million pounds), almonds (2.1 million pounds), grapes, (1.5 million pounds), grapefruit and apples (400,000 pounds each), and a variety of fruits, vegetables, and nuts.[7]

Since 1974, farmers have poured 4 billion pounds of Roundup on fields.[8] While independent studies of the safety of GE foods are scarce because the owners of the utility patents of GE plants refuse to allow truly independent research, there are many studies of Roundup and its main ingredient, glyphosate, that show it attacks the beneficial organisms in the human digestive system, causing serious health problems—increased birth defects, neurological developmental problems in children, kidney failure, respiratory problems, allergies.[9] Studies show that Roundup is a powerful soil biocide, resulting in the increase of microbial plant pathogens and mycotoxins.[10] The use of GE crops is part of the whole package of industrialized farming, an integrated system that enables corporate control of our food system (see fig. 1).

There has been a speedup in farming—a grain farmer used to be able to support a family with 160 acres; now it takes 1,600. The Roundup-GE package goes with the speedup. And U.S. government regulation has failed to protect the public from the toxic herbicides that farmers spray on GE crops.

One selling point of Roundup is that it breaks down quickly. That is why you can purchase it off the shelf in garden and hardware stores. That is accurate—it does break down quickly. But what Monsanto does not mention is that Roundup breaks down into AMPA, which lasts much longer and may be more toxic than glyphosate.[12]

Recently, the World Health Organization declared glyphosate a probable carcinogen, and many studies since 1985 have shown it to be an endocrine disruptor. But that did not stop the Environmental Protection Agency in 2013 from increasing the amount of glyphosate that is considered safe as a residue in soybeans, corn, and other crops.

It is extremely unscientific and poor public policy for the Patent and Trademark Office, a federal agency, to grant billions of dollars of utility patent rights for GE technology based on a demonstration of material difference, and for the Food and Drug Administration (FDA), another federal agency, to be simultaneously denying consumers basic information about the use of that technology in their food, based on that agency's finding of a lack of material difference.

Internal FDA documents obtained by the Alliance for Bio-Integrity during a 1998 lawsuit against the agency reveal that the FDA's biotechnology coordinator, James Maryanski, knew full well the potential risks but chose to override them. A November 1, 1991, memo to Maryanski titled "Points to Consider for Safety Evaluation of Genetically Modified Foods.

Supplemental Information" detailed the potential problems with new genetically engineered crops:

> increased levels of known naturally occurring toxicants, appearance of new, not previously identified toxicants, increased capability of concentrating toxic substances from the environment (e.g. pesticides or heavy metals) and undesirable alterations in the levels of nutrients.[13]

Despite this, the FDA claimed, and continues to assert, that genetically engineered foods are perfectly safe. Moreover, it also contends that they are "generally recognized as safe" (GRAS), even though none of them has met the requirements for GRAS status that are stipulated in the agency's own regulations.

The utility patents that Monsanto and other corporations hold on seed give them control over every use of that seed—including research to test it for safety. Farmers are not allowed to save and share the seed. University scientists must have permission from the patent holder and have to pay royalties to do research on patented seed. That is a significant barrier to independent science. By contrast, the Open Source Seed Initiative (OSSI) allows seed purchasers free use of the seed and asks for a pledge "not to restrict others' use of these seeds or their derivatives by patents or other means, and to include this pledge with any transfer of these seeds or their derivatives."[14]

Unfortunately, we do not live in a sweet world where researchers are free to work for the people and the earth and where decisions on what research gets funded are made purely on the basis of which projects benefit the largest number of poor and hungry people in former imperial colonies.

Just as the biotech proponents conflate "science" and GMOs, I and the organic movement conflate our struggle for family-scale, local, organic, or agroecological agriculture and against corporate control with the fight against GMOs. Buying Roundup Ready seed is not a free choice for farmers. Using that seed ensnares the farmer in Monsanto's clutches.

There are economic issues here as well. Major Goodman of North Carolina State University, a respected corn geneticist and member of the National Academy of Sciences, stated in testimony before the National Research Council that conventional breeding typically costs about $1 million per trait, while genetic engineering costs $136 million per GE trait, with most of the cost due to research and development, not to regulatory expenses.[15]

If we could assemble a council of farmers and scientists to evaluate the most cost-effective ways to invest public resources to eliminate world hunger, it is doubtful that they would choose genetic engineering. The 2008 United Nations International Assessment of Agricultural Knowledge, Science and Technology for Development (IAASTD), a comprehensive report on the future of farming, authored by four hundred scientists and backed by fifty-eight governments, stated that yields of GM crops were "highly variable" and in some cases "yields declined." The IAASTD concluded that since 70 percent of the world's population is fed by small farms, many run by women, the best course would be to increase investments in agroecology.[16]

That means doing the kind of work that the organic farmers of the Northeast, under the auspices of NOFA and MOFGA, have been doing for over four decades—helping family-scale farmers, homesteaders, and gardeners learn more about how to produce the healthiest, most nutrient-dense food using local resources, and providing ways for them to share this learning with one another. Farmer to farmer—campesino to campesino. Scientists who respect the indigenous knowledge of farmers can help the way Molly Jahns's team did in the Peacework pepper breeding project.[17]

If some of the millions spent on genetic engineering were directed to farmer-scientist organizations like the Farmer-Scientist Partnership for Development (MASIPAG) in the Philippines and the National Organic Agriculture Movement of Uganda (NOGAMU), there would be much more progress toward local self-reliance.

I am totally with the organic farmer who declared at our meeting that agriculture needs applied research and technology and that our land-grant universities should invest in applied research programs to find the solutions to the complex problems facing farmers today. The division of farmers into two camps—organic and conventional—is destructive and prevents us from learning from one another. It would help all farmers if the U.S. Department of Agriculture research apparatus had funding to support organic research programs at higher levels than the current barely 2 percent of all USDA research dollars.

So, at this time, I do not think the organic movement should drop our opposition to allowing certified organic farms to use GE crops. We should continue to campaign for labeling GMOs. We should call upon Cornell University to disband its propaganda arm for GMOs—the Alliance for Science—or repurpose it under the guidance of a board that represents the interests of the people of New York State. We should not oppose continuing research in genetic engineering, but we should demand that equal resources go into agroecology. Independent science is a great concept. Let's work together with conventional farmers for applied research that keeps our farms in business and allows us to grow the healthiest food for all the people while regenerating our soils and arresting climate change.

PS. I urge defenders of GMOs to read Steven M. Druker, *Altered Genes, Twisted Truth: How the Venture to Genetically Engineer Our Food Has Subverted Science, Corrupted Government, and Systematically Deceived the Public* (Salt Lake City: Clear River Press, 2015).

NOTES

1. "Gates Foundation Backed Pro-GMO Cornell Alliance for Science on the Attack," Corporate Crime Reporter, March 5, 2015, www.corporatecrimereporter.com/news/200/gates-foundation-backed-pro-gmo-cornell-alliance-science-attack.

2. Evelyn Fox Keller, *A Feeling for the Organism: The Life and Work of Barbara McClintock* (San Francisco: W.H. Freeman, 1983).

3. Jonathan Latham, "God's Red Pencil?: CRISPR and the Three Myths of Precise Genome Editing," *Independent Science News*, April 25, 2016, www.independentsciencenews.org/science-media/gods-red-pencil-crispr-and-the-three-myths-of-precise-genome-editing.

4. A. K. Wilson, J. R. Latham, and R. A. Steinbrecher. "Transformation-Induced Mutations in Transgenic Plants: Analysis and Biosafety Implications," *Biotechnology and Genetic Engineering Reviews* 23 (2006): 209–34.

5. M. V. Maffini, Heather M. Alger, Erik D. Olson,

and Thomas G. Neltner, "Looking Back to Look Forward: A Review of FDA's Food Additives Safety Assessment and Recommendations for Modernizing its Program," *Comprehensive Reviews in Food Science and Food Safety* 12 (2013), 439–53, doi:10.1111/1541-4337.12020.

6. D. Huber, "Letters from Prof. Don Huber to U.S. and EU Administrations Sent to President Jose-Manuel Barroso, EU President, cc to President Herman Van Rompuy, President Jerzy Buzek, Commissioner John Dalli, and some MEPs," March 25, 2011. These letters include extensive research citations which support this statement.

7. Carey Gillam, "What Is Going on with Glyphosate? EPA's Odd Handling of Controversial Chemical," *Huffington Post*, May 3, 2016, www.huffingtonpost.com /carey-gillam/what-is-going-on-with-gly_b_9825326.html.

8. C. Benbrook, "Trends in Glyphosate Herbicide Use in the United States and Globally," *Environmental Sciences Europe* 28:3 (2016), doi:10.1186/s12302-016-0070-0.

9. D. W. Sparling, C. Matson, J. Bickham, and P. Doelling-Brown, "Toxicity of Glyphosate as Glypro and LI700 to Red-eared Slider (*Trachemys scripta elegans*) Embryos and Early Hatchlings," *Environmental Toxicology and Chemistry* 25:10 (2006), 2768–74, doi:10.1897/05-152.1; D. Benedetti, E. Nunes, M. Sarmento, C. Porto, C. E. I. dos Santos, J. F. Dias, et al. "Genetic Damage in Soybean Workers Exposed to Pesticides: Evaluation with the Comet and Buccal Micronucleus Cytome Assays. Mutation Research 752 (2013), 28–33; S. L. Lopez, D. Aiassa, S. Benitez-Leite, R. Lajmanovich, F. Manas, G. Poletta, et al., "Pesticides Used in South American GMO-Based Agriculture: A Review of Their Effects on Humans and Animal Models," in *Advances in Molecular Toxicology*, edited by J. C. Fishbein and J. M. Heilman (New York: Elsevier, 2012), 41–75; R. Mesnage, M. Arno, M. Costanzo, M. Malatesta, G.-E. Séralini, and M. N. Antoniou "Transcriptome Profile Analysis Reflects Rat Liver and Kidney Damage Following Chronic Ultra-Low Dose Roundup Exposure," *Environmental Health* 14:70 (2015), doi:10.1186/s12940-015-0056-1; M. A. Marin-Morales, B. de Campos Ventura-Camargo, and M. M. Hoshina MM. "Toxicity of Herbicides: Impact on Aquatic and Soil Biota and Human Health. Herbicides–Current Research and Case Studies in Use," 2013, 399–443, http://cdn.intechopen. com/pdfs-wm/44984.pdf; Steeve Gress, SandrineLemoine, Gilles-Eric Séralini, and Paolo Emilio Puddu, "Glyphosate-Based Herbicides Potently Affect Cardiovascular System in Mammals: Review of the Literature," *Cardiovascular Toxicology* 15 (2014): 117–26, doi:10.1007/s12012-014-9282-y.

10. G. S. Johal and J. E. Rahe, "Glyphosate, Hypersensitivity and Phytoalexin Accumulation in the Incompatible Bean Anthracnose Host-Parasite Interaction," *Molecular Plant Pathology* 32 (1988), 267–81; M. R. Fernandez, F. Selles, D. Gehl, R. M. DePauw, and R. P. Zentner, "Crop Production Factors Associated with Fusarium Head Blight in Spring Wheat in Eastern Saskatchewan," Crop Science 45 (2005), 1908–16; Robert J. Kremer and Nathan E. Means, "Glyphosate and Glyphosate-Resistant Crop Interactions with Rhizosphere Microorganisms," European Journal of Agronomy 31 (2009), 153–61; G. S. Johal and D. Huber, "Glyphosate Effects on Diseases of Plants," *European Journal of Agronomy* 31 (2009), 144–52.

11. Cited in Benbrook, "Trends in Glyphosate Herbicide Use."

12. N. T. L. Torstensson, L. Lundgren, and J. Stenström, "InfluEnce of Climatic and Edaphic Factors on Persistence of Glyphosate and 2,4-D in Forest Soils," *Ecotoxicology and Environmental Safety* 18:2 (1989), 230–39, doi:10.1016/0147-6513(89)90084-5; M. M. Andréa, T. B. Peres, L.C. Luchini, S. Bazarin, S. Papini, M. B. Matallo, and V. L. Savoy, "Influence of Repeated Applications of Glyphosate on Its Persistence and Soil Bioactivity," *Pesquisa Agropecuária Brasileira* 38 (2003), 1329–35; W.A. Battaglin, M.T. Meyer, K.M. Kuivila, and J.E. Dietze, "Glyphosate and Its Degradation Product AMPA Occur Frequently and Widely in U.S. Soils, Surface Water, Groundwater, and Precipitation," *Journal of the American Water Resources Association* 50 (2014), 275–90.

13. Cited in Steven M. Druker, *Altered Genes, Twisted Truth: How the Venture to Genetically Engineer Our Food Has Subverted Science, Corrupted Government, and Systematically Deceived the Public* (Salt Lake City: Clear River Press, 2015).

14. FEDCO Seeds, 2016 Catalogue, 3.

15. The National Acadamies of Sciences, Engineering, and Medicine, "GE Crops: Meeting 1, Day 1, Major Goodman," 2014, http://vimeo.com/album/3051031 /video/106866601.

16. Jonathan Latham and Allison Wilson, "How the Science Media Failed the IAASTD," Independent Science News, April 7, 2008, www.independentsciencenews.org /environment/science-media-failed-the-iaastd/.

17. Michael Mazourek, George Moriarty, Michael Glos, Maryann Fink, and Mary Kreitinger, "Peacework: A Cucumber Mosaic Virus–Resistant Early Red Bell Pepper for Organic Systems," *HortScience*, January 13, 2009, http:// hortsci.ashspublications.org/content/44/5/1464.full.

Reprinted from Elizabeth Henderson, "Organic Farmers Are Not Anti-Science but Genetic Engineers Often Are," The Prying Mantis, May 2016, https://thepryingmantis .wordpress.com/.

Samuel H. Gottscho

ICE-MINUS BACTERIA

Ice-minus bacteria is a common name given to a variant of the common bacterium *Pseudomonas syringae*. This strain of *P. syringae* lacks the ability to produce a certain surface protein, usually found on wild-type *P. syringae*. The "ice-plus" protein, the "Ice nucleation-active" or Ina protein, found on the outer bacterial cell wall, acts as the nucleating centers for ice crystals. This facilitates ice formation, hence the designation "ice-plus." The ice-minus variant of *P. syringae* is a mutant, lacking the gene responsible for ice-nucleating surface protein production. This lack of surface protein provides a less favorable environment for ice formation. Both strains of *P. syringae* occur naturally, but recombinant DNA technology has allowed for the synthetic removal or alteration of specific genes, enabling the creation of the ice-minus strain.

The ice nucleating nature of *P. syringae* incites frost development, freezing the buds of the plant and destroying the occurring crop. The introduction of an ice-minus strain of *P. syringae* to the surface of plants would reduce the amount of ice nucleate present, rendering higher crop yields. The recombinant form was developed as a commercial product known as Frostban. Field-testing of Frostban was the first release of a genetically modified organism into the environment. The testing was very controversial and drove the formation of U.S. biotechnology policy. Frostban was never marketed.

From Wikipedia
https://en.wikipedia.org/wiki/Ice-minus_bacteria

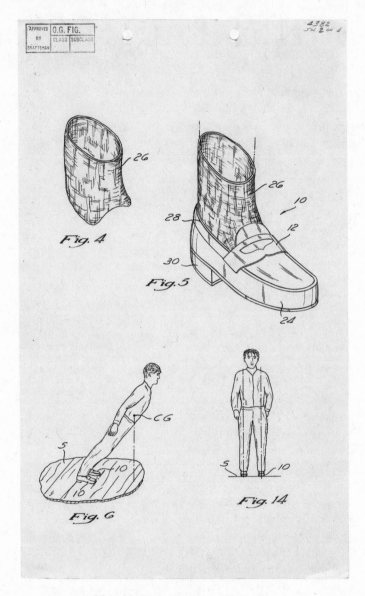

Michael Jackson's Anti-Gravity Illusion Shoes

Patent 5,255,452, Method and Means for Creating Anti-Gravity Illusion, Inventors: Michael J. Jackson; Michael L. Bush; and Dennis Tompkins, October 26, 1993

Records of the Patent and Trademark Office
National Archives and Records Administration

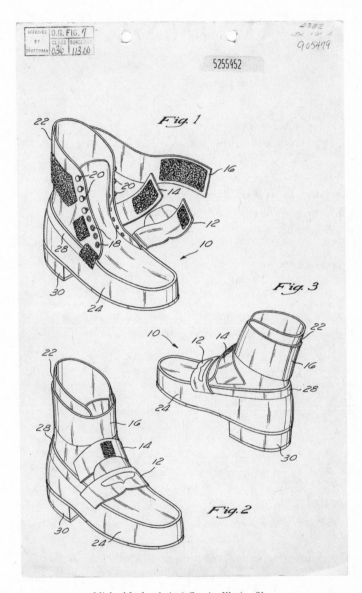

Michael Jackson's Anti-Gravity Illusion Shoes

Patent 5,255,452, Method and Means for Creating Anti-Gravity Illusion, Inventors: Michael J. Jackson; Michael L. Bush; and Dennis Tompkins, October 26, 1993

Records of the Patent and Trademark Office
National Archives and Records Administration

RECIPE FOR DISASTER

MARTIN POWELL

Forgive us Monsanto for being so bold
We were told that revenge is a dish best served cold ...

The seeds of corporate greed over centuries sown
Long term effects on us of which have yet to be known
You've always reaped the benefits of corporate clones
But now it's you Monsanto that we have outgrown

The lies of your insecticides are starting to flower
Like politicians planted in positions of power

Did you think that we'd forgotten all your government ties?
The rotten fruits of your labour, so ripe for disguise

Dine Vietnam
The Specials tonight: Feast your eyes!

Bad Eggs with Uncle Sam's "Orange Surprise"
Served in surgical strikes mainly launched from the skies

(Caution: May contain chemical agents and lies)

From generational genocide of children you slaughtered
To pesticides pushed upon our sons and daughters

Fast forward through time and we're asking ourselves
How the same toxins that made life in Vietnam hell
Are now active ingredients in the food that you sell?
Dioxin's boxed-in and on the way to our shelves?

R. de Salis

Asbestos-sheet-coated barn

When Monsanto modifies our federal officials
It's not just our food that becomes artificial
Collateral damage now so unilateral
They're selling us pests inside food labelled "natural"
And even made legal these dark arts they master
It's the wealthy's unhealthy recipe for disaster

But the alarm has been sounded; time to Roundup these villains
Cut the GMO crop of propped up politicians
Uproot the truth to dish the dirt on their unearthly decisions
And campaign against the grain of their botanical prison
Until organic revolution panics them to submission

And once we've baked all the fateful mistakes of the hateful
We'll serve sweet justice up in the States by the plateful

SELF PORTRAIT WITH WHITE BIRDS

LUCAS FARRELL

I stand surrounded
in the hayloft.
Beams of light
stream through
the myriad of bullet holes
in the siding.
I want to kill
nothing. With
precision.

HOOF TRIMMING

LUCAS FARRELL

Is like a day at the spa
for your goats. And not just
any spa. A torture spa.

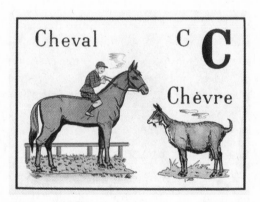

HALF OF ALL CHILDREN WILL BE AUTISTIC BY 2025, WARNS SENIOR RESEARCH SCIENTIST AT MIT

ALLIANCE FOR NATURAL HEALTH USA

Why? Evidence points to glyphosate toxicity from the overuse of Monsanto's Roundup herbicide on our food. For over three decades, Stephanie Seneff has researched biology and technology, over the years publishing over 170 scholarly peer-reviewed articles. In recent years she has concentrated on the relationship between nutrition and health, tackling such topics as Alzheimer's, autism, and cardiovascular diseases as well as the impact of nutritional deficiencies and environmental toxins on human health.

At a conference in December 2014, in a special panel discussion about genetically modified organisms (GMOs), she took the audience by surprise when she declared, "At today's rate, by 2025, one in two children will be autistic." She noted that the side effects of autism closely mimic those of glyphosate toxicity, and presented data showing a remarkably consistent correlation between the use of Roundup on crops (and the creation of Roundup-ready GMO crop seeds) with rising rates of autism.[1] Children with autism have biomarkers indicative of excessive glyphosate, including zinc and iron deficiency, low serum sulfate, seizures, and mitochondrial disorder.

A fellow panelist reported that after Seneff's presentation, "All of the seventy or so people in attendance were squirming, likely because they now had serious misgivings about serving their kids, or themselves, anything with corn or soy, which are nearly all genetically modified and thus tainted with Roundup and its glyphosate."[2]

Seneff noted the ubiquity of glyphosate's use. Because it is used on corn and soy, all soft drinks and candies sweetened with corn syrup and all chips and cereals that contain soy fillers have small amounts of glyphosate in them, as do our beef and poultry, since cattle and chicken are fed GMO corn or soy. Wheat is often sprayed with Roundup just prior to being harvested, which means that all non-organic bread and wheat products are also sources of glyphosate toxicity. The amount of glyphosate in each product may not be large, but the cumulative effect, especially with as much processed food as Americans eat, could be devastating. A recent study shows that pregnant women living near farms where pesticides are applied have a 60 percent increased risk of children having an autism spectrum disorder.[3]

Other toxic substances may also be autism-inducing. The Alliance for Natural Health USA (ANH) published a story on the Centers for Disease Control and Prevention whistleblower who revealed the government's deliberate concealment of the link between the MMR vaccine (for measles, mumps, and rubella) and a sharply increased risk of autism, particularly in African American boys.[4] Other studies now show a link between children's exposure to pesticides and autism.[5] Children who live in homes with vinyl floors, which can emit phthalate chemicals, are more likely to have autism. Children whose mothers smoked were also twice as likely to have autism. Research now acknowledges that environmental contaminants such as PCBs, PBDEs, and mercury can alter brain neuron functioning even before a child is born.

In 2014 the U.S. Department of Agriculture (USDA) released a study finding that although there were detectable levels of pesticide residue in more than half of food tested by the agency, ninety-nine percent of samples taken were found to be within levels the government deems safe,

and forty percent were found to have no detectable trace of pesticides at all. The USDA added, however, that due to "cost concerns," it did not test for residues of glyphosate.[6] Let's repeat that: they never tested for the active ingredient in the most widely used herbicide in the world. "Cost concerns"? How absurd—unless they mean it will cost them too much in terms of the special relationship between the USDA and Monsanto. You may recall the revolving door between Monsanto and the federal government, with agency officials becoming high-paying executives and vice versa. Money, power, prestige: it's all there. Monsanto and the USDA love to scratch each other's backs. Clearly this omission was purposeful.

In addition, the ANH reported that the number of adverse reactions from vaccines can be correlated as well with autism,[7] though Seneff says it doesn't correlate quite as closely as with Roundup. The same correlations between applications of glyphosate and autism show up in deaths from senility.

Of course, autism is a complex problem with many potential causes. Seneff's data, however, is particularly important considering how close the correlation is—and because it is coming from a scientist with impeccable credentials. In 2014 she spoke at the Autism One conference and presented many of the same facts; that presentation is available on YouTube.[8]

Monsanto claims that Roundup is harmless to humans. Bacteria, fungi, algae, parasites, and plants use a seven-step metabolic route known as the shikimate pathway for the biosynthesis of aromatic amino acids; glyphosate inhibits this pathway, causing the plant to die, which is why it's so effective as an herbicide.[9] Monsanto says humans don't have this shikimate pathway, so it's perfectly safe.

Seneff points out, however, that our gut bacteria do have this pathway, and that's crucial because these bacteria supply our body with crucial amino acids. Roundup thus kills beneficial gut bacteria, allowing pathogens to grow; interferes with the synthesis of amino acids, including methionine, which leads to shortages in critical neurotransmitters and folate; removes important minerals like iron, cobalt, and manganese; and much more.

Even worse, she notes, additional chemicals in Roundup are untested because they're classified as inert, yet according to a 2014 study in BioMed Research International, these chemicals are capable of amplifying the toxic effects of Roundup hundreds of times over.[10]

Glyphosate is present in unusually high quantities in the breast milk of American mothers, at anywhere from 760 to 1,600 times the allowable limits in European drinking water. Urine testing shows Americans have ten times the glyphosate accumulation as Europeans. "In my view, the situation is almost beyond repair," Seneff said after her presentation. "We need to do something drastic."

.davidgumpert.com/no-mr-nice-guy-food-club-members -seek-uphold-food-rights.

3. "Autism-Pesticide Link Found in California Study," Seeker, June 23, 2014, www.seeker.com/autism-pesticide -link-found-in-calif-study-1768739610.html.

4. Alliance For Natural Health USA, "Autism -Vaccine Cover-up Snowballs as Whistleblower's Identity Is Revealed," August 26, 2014, www.anh-usa.org/autism -vaccine-cover-up-latest-updates.

5. Alliance For Natural Health USA, "Genetics + Environmental Chemical Soup = Autism?" April 14, 2009, www.anh-usa.org/genetics-environmental-chemical-soup -autism.

6. Carey Gillam, "USDA Report Says Pesticide Residues in Food Nothing to Fear," Reuters Health News, December 19, 2014, www.reuters.com/article/us-usda -pesticides-report-idUSKBN0JX2FZ20141219.

7. Alliance For Natural Health USA, "Autism Diagnoses Have Risen by 78% over the Last Decade," April 10, 2012, www.anh-usa.org/autism-diagnoses-have-risen.

8. Stephanie Seneff, "Autism Explained: Synergistic Poisoning from Aluminum and Glyphosate," August 13, 2014, https://www.youtube.com/watch?v=a52vAx9HaCI.

9. "Monsanto's Roundup Herbicide May Be Most Important Factor in Development of Autism and Other Chronic Disease," Mercola, June 9, 2013, http://articles. mercola.com/sites/articles/archive/2013/06/09/monsanto -roundup-herbicide.aspx.

10. Robin Mesnage, Nicolas Defarge, Joël Spiroux de Vendômois, and Gilles-Eric Séralini, "Major Pesticides Are More Toxic to Human Cells Than Their Declared Active Principles," BioMed Research International 2014, doi:10.1155/2014/179691.

From the Alliance For Natural Health USA website, www.anh-usa.org. Reprinted with permission.

NOTES

1. Nick Meyer, "MIT Researcher's New Warning: At Today's Rate, Half of All U.S. Children Will Be Autistic by 2025," The Mind Unleashed, October 28, 2014, http:// themindunleashed.com/2014/10/mit-researchers-new -warning-todays-rate-half-u-s-children-will-autistic-2025 .html.

2. David Gumpert, "No More Mr. Nice Guy: Food Club Members Seek to Uphold Food Rights," The Complete Patient, September 24, 2016, www

DECEMBER

INTERGENERATIONAL DYNAMICS AND HISTORY

R. de Salis

Ancient Yew

FROM *CORPS*

PHIL CORDELLI

The heart's a hollow organ
grown around chambers, twitching

cutworm curled up
eating cornmeal

the tree is prepared
like a child from another time

tanks of brittle foam
that in so dense a fizz

I place the stem, plunge, watch the circular dial
diagonally lined

through the rain one thinks
thin organs of thought

pocked, and in the cavities
a vision of all awaiting comedy

in the temple, in the town, in the field

matted roots of the upturned tomato
sound of the fourwheeler

Love and loss coexist
begin to tinge upon each other

constant sticks of wilderness
in the peaks above

actually the opposite of permanence

THE WINDS

DOUGLASS DECANDIA

When the strong wind comes
and scatters
all the things
we have been holding

Lift your vision to the trees,
to the branches
who once, too
held golden leaves

And see, like it was
the first time,
the empty forest
filling with light.

R. de Salis

Ash Tree Regrowth

LANDSCAPE AND IDENTITY

LAURA DEL CAMPO

Gisela Weimann

My point is that food is a cultural product; it cannot be produced by technology alone.… A culture is not a collection of relics or ornaments, but a practical necessity, and its corruption invokes calamity. A healthy culture is a communal order of memory, insight, value, work, conviviality, reverence, aspiration. It reveals human necessities and the human limits. It clarifies our inescapable bonds to the earth and to each other. It assures that the necessary restraints are observed, that the necessary work is done, and that it is done well.
—Wendell Berry

When my mother bought this property, before anything was built, she walked the land. She would visit often, park her car in front of a gravel mound where the driveway abruptly ended, and walk. From December to July she visited in all types of weather. With each visit, time on the land allowed her to see the farm and her family in an ecological context. The farm's place in the fragile mosaic of the Sourland Mountains became more and more obvious. By the time the house was built, she knew it rested on the highest point on the property. She knew the soil type was clay loam and that the wetlands would develop on the northwestern slope of the property. She knew how the deer moved across and at which times; she knew where each fox den rested.

The landscape in which we live is an inextricable part of our identity. It is the subtle and the grand—the spheres of interaction between communities of species, geologic terrain, and the more general bioregion. Our chosen agricultural system has the ability to either support or fragment our landscape. The technology we use and the manner by which we engage with our bioregion is a choice, one paramount in developing a sense of connection and a sense of pride.

Our food system speaks of our community, our identity within a landscape, our symbolic and physical relation to place. Transparency is necessary for landscape to be consciously recognized as an integral component of both our personal identity and communal narrative. Transparency of food production enables people to see how consumptive patterns affect our quality of life within landscape—to understand the true costs of disparate agricultural systems. It demystifies the qualitative gap between independent local production and corporate agribusiness. As we

make obvious the interconnection of relationships within a system, we become capable of grasping the narrative of the very composition of our bodies, whether that narrative is one of communion or commodity.

The end products of our food systems are, of course, not isolated from the conditions within which they were produced, though often we experience them this way. Food processing, packaging, and marketing serve to conceal the origins of food. Yet it is the visceral experience of the production chain, an unveiling of our food's journey and so the revealing of our own story, that enables people to experience the connection between their own body and health as it relates to the natural abundance of the land.

When we are involved with the management of our agriculture systems, we are necessarily involved with how our land is developed. Higher reliance on local production and resources affords less cause to assimilate land-use patterns to world market standards. With such involvement, communities have more control of local resources and therefore more freedom to use land in a way suitable to their heritage, the legacy of the land, and the ecology of their bioregion.

Communities can improve natural abundance by adapting agriculture to reflect ecological systems and by consciously sustaining local nuance, traditional practices, and technologies that serve cultural identity within a landscape. Natural abundance increases local autonomy, allowing communities to renounce sole dependence on the global market. The stronger human culture is united within an ecological community, the more cultural practice evolves to reflect the provisions of the local landscape. We are, after all, not building these systems for ourselves but for our children.

A SHORT, UNHAPPY HISTORY OF BUSINESS ADVICE FOR FARMERS:

Animadversions on Market Mavins, Together With Some Word of Counsel for Young Farmers

MICHAEL FOLEY

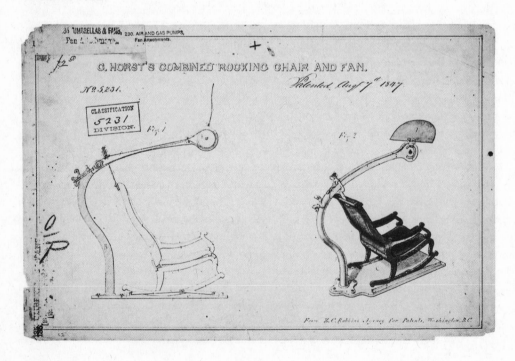

In the 1920s the newly minted profession of agricultural economist plied its wares among America's struggling family farmers. The farmers had profited from cheap land, thanks to the Homestead Act, and high commodity prices, thanks to World War I. In 1918 commodity prices collapsed. In the upper Plains states, recently settled by waves of homesteaders, including my great-grandmother, the problem was compounded by the beginnings of what became the Dust Bowl, also starting in 1918 with a drought year soon to be followed by many.

Savvy economists and their friends in banking concluded that the failure of so many family farms lay in the farmers' lack of business sense. Extension agents, trained by university-bred agricultural economists, traveled the country teaching farmers how to keep books, analyze a balance sheet, and choose wisely among competing crop options. They also pushed mechanization, encouraging farmers to expand and adopt the newly developed tractors in place of horse-drawn agricultural implements. Not many farmers bought. They knew horses. They didn't know the new machines, and they were all too aware of how unreliable they were. Besides, buying a tractor or combine required borrowing. The bankers and economists were eager to counter the shibboleth against indebtedness, but most farmers were not interested.

The tractor companies mostly moved offshore, selling more tractors in the Soviet Union, where the government was eager to apply the latest American technology, than at home during the twenties. Some of the extension agents went to work for the Soviets, too, helping them devise the enormous state farms that they, along with their Communist sponsors, thought the only solution to farming on a modern scale. A few such experiments were tried in the United States, without the coercive powers of the State to back them, but they almost uniformly failed.

For the "progressive" farmers who bought into the new economic logic, there was plenty of land to expand into, as homesteaders encouraged by agricultural experts and politicians to strike out West with the promise that "rain follows the plow," went under in the growing drought conditions of the twenties. Many of the tractor operators failed, too, and it wasn't until World War II brought higher commodity prices that farmers once again flourished.

Most had long since fled the countryside, like great-grandmother Sarah McCann, who homesteaded outside of Lewiston, Montana. As the farm economy failed, she moved to what was still a boom town, and then to another, Roundup, Montana, where my father was born. But no one remains from that clan. All moved on. Lewiston has recovered a little of it's former wealth. Roundup still struggles.

The take-away for farmers: if you depend on the "experts" for advice, some goddamned fool is likely to make a goddamned fool out of you. The first fool will keep his well-paying job at the university or bank. You'll be left to pack up and find another way to make a livelihood for yourself and your family. Thus was the American countryside emptied.

The emptying, of course, would take a while longer. Following World War II, Americans finally bought the promise of the new farm machinery. Many young men from farming families had learned how to handle heavy equipment in the battlefields of Europe. Those who stayed home had tinkered with their autos long enough to feel comfortable with the internal combustion engine. And farm machinery was cheap. So was credit. Farmers expanded, at first slowly, and then, as farm prices continued to fluctuate, driving indebted farmers from ancestral homes, more rapidly on the newly freed land.

The logic of the ag economists, already evident in the twenties, fifty years before Secretary of Agriculture Earl Butz articulated it clearly, was "Get big or get out." But getting big meant doing your cost accounting, identifying the profitable crop or two, and putting everything into it. In the course of the sixties and especially the seventies, when the deficiencies of Soviet agriculture created a boom market for American grain, farmers took out fences and hedgerows, cut the woodlot,

let the vegetable garden go to grass, drove out the chickens and pigs and family cow, and tore down all the outbuildings that went with them. All those "uneconomical" ventures had to make way for rows upon rows upon rows of corn or soybeans, wheat or oats. Depopulated of farmers, the land was now depopulated of farm animals. Farm wives learned to buy their food at the new supermarkets and look for supplemental income, increasingly necessary to sustain a middle-class

farm lifestyle, hopefully with a good government job. No more churning butter and collecting eggs for sale in town. No more growing and canning food for the winter in the family vegetable garden. No more wood stoves fueled with farmstead wood. Fossil fuels were more "economical."

The new economies didn't avail much. The next downturn in commodity prices came in 1979, the same year that the certified economic genius Federal Reserve Chair Paul Volcker jacked up American interest rates, wrecking the world economy. Farms failed in the tens of thousands, saddled with debt trying to get big to avoid getting out. Between 1979 and 1985, fully half of American farms went under. American farmers had the options of watching the sheriff foreclose on the family farm, shooting their bankers, or shooting themselves. Most chose the first option. A few chose one or both of the latter. In any case, farmers were the losers. But they were doing what the experts said. They had gotten bigger; now they were getting out.

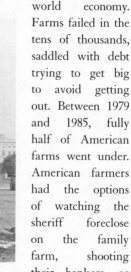

The story was, if anything, more tragic in what we used to call the Third World. The "green revolution" propagated by American crop breeders, the Rockefeller Foundation, and the

World Bank encouraged farmers in Mexico, India, the Philippines, and elsewhere to adopt new, highly productive crops developed by American researchers and their colleagues around the world. The new crops only needed abundant and reliable sources of water, fertilizer, and pesticides. And their introduction indeed rewarded the hopes of their promoters. Production of basic commodities, principally rice, wheat, and corn, soared in adopting countries. The progressive farmers who seized on the opportunities promised were lauded by agricultural extension agents and their governments and celebrated in the press.

But the boom didn't last. Rapidly increased production drove down commodity prices, making the expense of the new techniques increasingly hard to manage for the average farmer, and many poorer farmers never participated. In Punjab and Rajasthan in northern India, aquifers were pumped dry, traditional waterworks left to decay, and much of the land desertified by a combination of soil-depleting fertilizers and drought. Thousands of farmers have committed suicide drinking the pesticides they no longer need on their parched land.

In Mexico, peasant farmers were coerced into adopting the new crops by monopoly buyers and rewarded with cheap fertilizer in exchange for political loyalty to the regime that promoted the new technology. Already by the 1980s, peasants reported that la tierra ya no da, "the land is exhausted," and they rightly pointed to the artificial fertilizers to which they had become

addicted. The worst blow came in the mid-1990s when NAFTA, promised by the economists to bring prosperity to Mexico, ushered in a flood of cheap U.S. corn, undermining the little cash flow that Mexican peasants depended on. Millions crossed the border, where cheap labor continues to suppress the prices of the fruits and vegetables upon which a new generation of farmers is hoping to depend.

Fast forward not so many years and we find a new

movement back to the farm, invited once again to acquire the business skills without which, so all their would-be advisers insist, they cannot but fail. One of the most popular recent books on the subject, Richard Wiswall's The Organic Farmer's Business Handbook, at least comes from a market gardener—a farmer, that is—who appreciates a diverse cropping and marketing strategy. But Wiswall echoes the cost-accounting advice of the most simpleminded ag economists of the 1920s.

Let's do a "crop enterprise budget," crop by crop, Wiswall advises, so we know which crops are worth growing, which not. Presumably, Wiswall is aware that a diversified strategy can protect the farmer against the failure of a given crop. He is not headed down the path of monoculture. But his economic logic is the same that brought us to the current state of American agriculture. And there is no mention in Wiswall of the subsistence strategies—the woodlot, a flock of chickens, a couple of pigs, a dairy cow or goats, and cooperative labor arrangements—which many of the new farmers have adopted for reasons that their would-be advisers inevitably dismiss as "romantic." If it can't be monetized, it can't be counted, and doesn't then count.

Is there an alternative? Certainly it's a good thing to keep the books, minimize expenses and maximize sales. It can't hurt, either, to be aware of the profit margins available for different crops and "enterprises" on the farm. But are we condemned to the sterile—and too often disastrous—logic of cost-accounting?

The Russian agrarian sociologist and economist Alexander Chayanov found that peasant economies were based on a more complex economic logic. Though many peasant farmers around the world produce for the market, they also produce for themselves. Their production expands and their income rises, not as their business sense increases but as their families grow. And once grown and provided for, peasant farmers scale back. In some cases, subsequent studies have found, especially where market sales don't provide a reliable income from year to year, their attitude is "subsistence first," meaning they will reject any agricultural innovation, including

supposedly more lucrative crops, that threatens proven strategies of providing for their families. Chayanov was executed by Stalin in the 1930s for insisting that, given that logic, the large-scale collective and state farms Stalin favored would fail (and fail they did). But communists and capitalists shared their disdain for the peasant household economy. The peasant "subsistence first" attitude has driven agricultural advisers to distraction, from British and French colonial agents to the U.S. Agency for International Development advisers and even Peace Corps volunteers.

We all have to pay the bills, no doubt. But the woodlot, vegetables, and fruit from our market gardens, that flock of chickens or ducks, those pigs and milk animals are also part of the value calculation. It makes sense to cling to them in a market-driven world, because they provide the security and resilience that the market cannot. It also makes sense, of course, to compare the costs of maintaining animals with the costs of the same goods bought at market; but we all know that our chickens' eggs are much more healthful than the supermarket variety, "free-range," "organic," or not. These aren't trivial considerations, as the present state of American farming and health demonstrate.

There is something else missing from the cost-accounting of the business experts. A character in one of Wendell Berry's novels recalls working on every farm down his road, and he's proud to say that not a penny was exchanged. Cooperative labor arrangements were common in American farming almost to the mid-twentieth century, before large-scale fruit and vegetable production on the California model became common. Harvest time brought out everyone— farmers and their workers, farm wives, town wives and their children, even college kids home from school for the weekend. Orchards were harvested, tobacco sorted, tomatoes brought in. School children would take a couple of weeks off in the Santa Rosa area just for the prune harvest.

The newest farmers, and most of the smaller farm owners of all ages, rely on volunteer labor even today. Collective work parties are part of the new ethic of solidarity among young farmers. Sharing resources—traditional in the older American agriculture—is embodied in the structure of the Greenhorns and the Farmers Guild. And sharing knowledge is part of the farmer and farm intern exchanges of the CRAFT movement. Our business advisers would prefer we become employers and leave the schooling to accredited colleges and universities. Barter? What's that?

The values that small farmers often embody are outside the logic of cost-accounting. They recall a time and a society that was more resilient than ours, did a better job in some cases of taking care of its own, and that provided for most everyone in hard times and good. Not a perfect one, to be sure, but one where frugality, self-sufficiency, and community embodied the wisdom of the Chayanovian peasant, teaching that we will only survive the vagaries of the market if we provide prudently for ourselves and share with others.

We can cost out everything and still come up losers. The winners will always be the bankers, tractor manufacturers, chemical and seed companies, and, of course, that legion of agricultural advisers, including business advisers, so willing to help us out. For farmers "acquiring business skills" has too often left us with debt, diabetes, and the death of our dreams. It's time to learn to take all that advice with a grain of salt. Otherwise, some damned fool will make a damned fool out of you, keeping his or her well-paying job (even better paid, in relation to farmers, than in the twenties), while you limp off to find another way of life and another way to provide a livelihood for yourself and your family.

AN OLD PEASANT'S ADVICE FOR YOUNG FARMERS

The economists should have known better than to throw farmers on the mercies of the market. Farmers, they all acknowledge, are price takers, not price makers. We are at the market's mercy. But the economists believe in a self-equilibrating model of market behavior. Alas, the economists' god gives no thought to the wrecked lives, lost topsoil, or abandoned communities the market leaves behind in the process of reaching a new equilibrium. All those lost farmers are just so much fodder for the efficient market game.

You need to know better, however. So here are some rough rules of thumb for new and old farmers:

1. If you have no room left for anything but market crops, or are too busy meeting market demands to provide for a substantial part of your own subsistence, you're in trouble. At the very least, your health and that of your family may be endangered by all that store-bought food you've let yourself rely on.

2. If you're so dependent on operating loans that a significant market downturn or crop failure will make paying the bills (never mind the bank) difficult, you're over-leveraged. The market will take a significant downturn, if not this year, then soon enough. You will have crop failures.

3. If you've expanded to the hilt and found just the formula to balance income and expenses, expect to fail. The market will cheat you, and the income will not cover the expenses, and you'll have to find some way to swim in a rising tide, or you'll sink, like too many of your predecessors.

4. If you don't have rich ties of mutual help and exchange with other people in your community, you're endangering both them and you. Build those ties through labor exchanges, loans, barter, and fun.

5. Diversify. If you owe money on the farm and money on the equipment and money on student loans, you can't afford to rely on one or two crops or one or two markets. Those crops that were worth a million last year may be worth next to nothing when the natural foods market finds a bunch of new suppliers.

6. Above all, curb your appetites. You can "live like a pauper, eat like a king" and be surrounded by good neighbors and a resilient community. Or you can pursue the consumer dreams of those who believe they can always rely on the next paycheck and their IRA.

7. You will never save enough to pay for your child's college education or avoid mortgaging the farm to pay what the college loan racketeers consider the "family share" of the bill. Better to put the farm in a land trust, get rid of all your assets, and count on a free ride for the kid.

8. Just don't build that nice big house. Don't buy the latest, greatest, GPS-driven tractor and computer-driven fertilizer-cultivator-drill attachment. When the market collapses and the sheriff comes to foreclose, he'll want the keys to the house and the tractor, the barn, the laser-leveled field, the whole shebang.

9. Avoid the sheriff.

10. Avoid the banker too.

11. Embrace your neighbor. Build community. Bring people together.

Mynah birds

AGAINST FORGETTING

DERRICK JENSEN

Last night a host of nonhuman neighbors paid me a visit. First, two gray foxes sauntered up, including an older female who lost her tail to a leghold trap six or seven years ago. They trotted back into a thicker part of the forest, and a few minutes later a raccoon ambled forward. After he left I saw the two foxes again. Later, they went around the right side of a redwood tree as a black bear approached around the left. He sat on the porch for a while, and then walked off into the night. Then the foxes returned, hung out, and, when I looked away for a moment then looked back, they were gone. It wasn't too long before the bear returned to lie on the porch. After a brief nap, he went away. The raccoon came back and brought two friends. When they left, the foxes returned, and after the foxes came the bear. The evening was like a French farce: as one character exited stage left, another entered stage right.

Although I see some of these nonhuman neighbors daily, I was entranced and delighted to see so many of them over the span of just one evening. I remained delighted until sometime the next day, when I remembered reading that, prior to conquest by the Europeans, people in this region could expect to see a grizzly bear every fifteen minutes.

This phenomenon is something we all encounter daily, even if some of us rarely notice it.

It happens often enough to have a name: declining baselines. The phrase describes the process of becoming accustomed to and accepting as normal worsening conditions. Along with normalization can come a forgetting that things were not always this way. And this can lead to further acceptance and further normalization, which leads to further amnesia, and so on. Meanwhile the world is killed, species by species, biome by biome. And we are happy when we see the ever-dwindling number of survivors.

I've gone on the salmon-spawning tours that local environmentalists give, and I'm not the only person who by the end is openly weeping. If we're lucky, we see fifteen fish. Prior to conquest there were so many fish the rivers were described as "black and roiling." And it's not just salmon. Only five years ago, whenever I'd pick up a piece of firewood, I'd have to take off a half-dozen sowbugs. It's taken me all winter this year to see as many. And I used to go on spider patrol before I took a shower, in order to remove them to safety before the deluge. I still go on spider patrol, but now it's mostly pro forma. The spiders are gone. My mother used to put up five hummingbird feeders, and the birds would fight over those. Now she puts up two, and as often as not the sugar ferments before anyone eats it. I used to routinely see bats in the summer. Last year I saw one.

You can transpose this story to wherever you live and whatever members of the nonhuman community live there with you. I was horrified a few years ago to read that many songbird populations on the Atlantic Seaboard have collapsed by up to 80 percent over the last 40 years. But, and this is precisely the point, I was even more horrified when I realized that Silent Spring came out more than forty years ago, so this 80 percent decline followed an already huge decline caused by pesticides, which followed another undoubtedly huge decline caused by the deforestation, conversion to agriculture, and urbanization that followed conquest.

My great-grandmother grew up in a sod house in Nebraska. When she was a tiny girl—in other words, only four human generations ago—there were still enough wild bison on the Plains that she was afraid lightning storms would spook them and they would trample her home. Who in Nebraska today worries about being trampled by bison? For that matter, who in Nebraska today even thinks about bison on a monthly, much less daily, basis?

This state of affairs is problematic for many reasons, not the least of which is that it's harder to fight for what you don't love than for what you do, and it's hard to love what you don't know you're missing. It's harder still to fight an injustice you do not perceive as an injustice but rather as just the way things are. How can you fight an injustice you never think about because it never occurs to you that things have ever been any different?

Declining baselines apply not only to the environment but to many fields. Take surveillance. Back in the 1930s, there were people who freaked out at the notion of being assigned a Social Security number, as it was "a number that will follow you from cradle to grave." But since 9/11, according to former National Security Agency official William Binney, the U.S. government has been retaining every email sent, in case any of us ever does anything the government doesn't like. How many people complain about that? And it's not just the government. I received spam birthday greetings this year from all sorts of commercial websites. How and why does ESPN.com have my birth date? And remember the fight about GMOs? They were perceived as scary (because they are), and now they're all over the place, but most people don't know or don't care. The same goes for nanotechnology.

Yesterday I ate a strawberry. Or rather, I ate a strawberry-shaped object that didn't have much taste. When did we stop noticing that strawberries, plums, and tomatoes no longer taste like what they resemble? In my twenties I rented a house where a previous resident's cat had pooped all over the dirt basement, which happened to be where the air intakes for the furnace were located. The house smelled like cat feces. After I'd been there a few months, I wrote to a friend, "At first the smell really got to me, but then, as with everything, I got used to the stench and it just doesn't bother me anymore."

This is a process we need to stop. Milan Kundera famously wrote, "The struggle of man against power is the struggle of memory against forgetting." Everything in this culture is aimed at helping to distract us from—or better, help us to forget—the injustices, the pain. And it is completely normal for us to want to be distracted from or to forget pain. Pain hurts. Which is why, on every level, from somatic reflex to socially constructed means of denial, we have pathways to avoid it.

But here is what I want you to do: I want you to go outside. I want you to listen to the (disappearing) frogs, to watch the (disappearing) fireflies. Even if you're in a city—especially if you're in a city—I want you to picture the land as it was before the land was built over. I want you to research who lived there. I want you to feel how it was then, feel how it wants to be. I want you to begin keeping a calendar of who you see and when: the first day each year you see buttercups, the first day frogs start singing, the last day you see robins in the fall, the first day for grasshoppers. In short, I want you to pay attention.

If you do this, your baseline will stop declining, because you'll have a record of what's being lost.

Do not go numb in the face of this data. Do not turn away. I want you to feel the pain. Keep it like a coal inside your coat, a coal that burns and burns. I want all of us to do this, because we should all want the pain of injustice to stop. We should want this pain to stop not because we get used to it and it just doesn't bother us anymore, but because we stop the injustices and destruction that are causing the pain in the first place. I want us to feel how awful the destruction is, and then act from this feeling.

And I promise you two things. One: feeling this pain won't kill you. And two: not feeling this pain, continuing to go numb and avoid it, will.

Originally published as Derrick Jensen, "Against Forgetting," *Orion*, July–August 2013.

Katsushika Hokusai

PROSPECT OF LONGEVITY

ELIZABETH KOHLER

Recreating history
not like reanimating
a corpse with voltage

but holding a mirror to it
tracing selected lines on transparent paper
amalgamating a body concealing life in it

like a small, battered animal
knowing hands, which build machines,
only as vengeful gods.

She divulges patience to it
with fishmeal from her own cauldron
twine from the hair of her dead horses

she discovers the uncertainty
of this body's furtive life
as water reflecting oil stains in the sky

or an unlocked chest, marooned at the shore
emptying itself of what carried it there.

It is not a yielding discovery,
which reassures the inconsolable
who surround the body
on the operating table of stainless steel

the mud on their faces
mimicking the body's whispers of finite

under the scattered drone
of stunned hummingbirds

on the wrong side of the glass.

Every trial holds a lesson
where they try to train the body

as if earth, or death, were a dog
needing familiarity
with workings untranslatable
to dog language.

She is numb to the prospect of longevity
while, over tea from the thriving herb garden,
they attempt to resolve its worth.

How do you amend what is disruptive by nature?
She demonstrates simplicity, crouched in a sun hat

observing a honeybee
copulate with a foxglove

uneasy at their justifications
for mechanized components.

She longs to gouge their unwillingness
to offer their own limbs

to rediscover this body
not as a dying fire
needing more bad fodder

or their reluctance
to compromise delusions of forever.

At nightfall, she sits in her plot of tenacious mint
chewing on their flowers, filling her head

with her own muttering
to the stolen limb of a walking tractor:

never command the moon

like a child in her angel costume
drowning a pale bird in its bath

Vincent van Gogh

Wheat Field

WORKSONGS

BENNETT KONESNI

It is a gray November morning and my fingers are numb from pulling the last of this year's root crop from the ground. I'm ready to quit when someone fires up a song. I join in, and so does the rest of the team.

Almost on their own, the parsnips pop out of the ground and into neat piles, ready for bins. The bins appear and the roots almost stack themselves, ready for the truck. Up they go into the bed, and as the truck rolls off I'm left with the crew for one last verse, which we holler out into the dusk as we walk to the barn. I realize that even though I'm sore from a day of work, I have more energy now than I had an hour ago.

This is what it feels like when we're in a groove, belting out worksongs in the field. Productive. Fun. It's playful, and yet things are getting done. And songs like this have a long history in Maine.

I first got into worksongs in 1995. I grew up here and always wanted to work aboard the local schooners off the coast. When I was thirteen I joined a daysailer out of Camden, and over the next five years I worked my way up to the Rockland windjammer the *J. & E. Riggin*. The *Riggin* takes weeklong sails out through

Penobscot Bay, and the captains of the boat, John Finger and Annie Mahle, have held onto a tradition of shipboard music that stretches back into the early 1800s, if not earlier.

Some mornings as they're bringing up the anchor, and many evenings as the day is winding down, the skippers and their crew break out old songs and share them with their passengers:

songs they learned as youngsters starting out on other boats in the bay, or songs they've picked up elsewhere over the years.

Some of the songs are sea shanties, musical tools that help people raise the sails and haul up the anchor more easily and in rhythm. Others are lyrical ballads that tell local stories or name islands, landmarks, whales. In either case, the music helps get the job done.

Some of my favorite shanties are those that were collected by the remarkable Joanna Colcord of Searsport, Maine, the next town over from Belfast. Joanna grew up on tall ships captained by her father, part of a long line of skippers of Maine schooners. Her book *Roll and Go: Songs of American Sailormen* has a collection of great songs and a stirring introduction that pointedly reflects on how changes in technology have ended most shanty-singing onboard ships, changing the nature and character of the people who do the work. I think about it often and wonder how technology has changed farmers, their songs, and their character.

When I started farming in 2000, the old sea songs turned out to be quite handy. We were wheel-hoeing a row of beans when an apprentice one row over shouted "Hey, Bennett, didn't you used to sing while doing hard rhythmic work on the boats?" Out of breath, I gasped, "Yep." "Well, why aren't we singing now?"

And so we launched into "Haul Away, Joe," and I began to wonder what songs old farmers used to sing to help get stuff done. It has led me on an odyssey that has taken me around the world and dropped me off back home a few miles from where I started.

It turns out worksongs are one of the secret

tools that the old-timers used in cultures all over the world. And here's why: they are efficient; they help you work harder, faster, longer. They are value added: they take a raw product or process and make it more valuable to you and your customers. And they have low overhead: you already have the equipment necessary.

What's more, they tie you to a place and a tradition, much like heritage breeds and seeds. Certain songs are useful only for certain jobs

or certain times of the year; others are more generally helpful but perhaps less effective than a purpose-built song. And in their surprising utility and beauty, they can tie a farm crew and its community together with enduring and mutually profitable bonds.

Before Europeans arrived, the Algonquian tribes who lived in Maine for several thousand years had songs for all manner of work, including paddling, hunting, and farming. I have been told of songs specifically for working with corn. The local Penobscot people had a "Green Corn Dance" that could easily have been sung in the fields and then performed at the first harvest of the green corn. This would fit a pattern of using songs for work and ceremony that can be found in many cultures.

These days the sea shanties are the best-known of Maine's worksong traditions, but they have a cousin in the "shanty songs" from the lumberjacks, which generally tell the stories of the trees, rivers, log jams, and colorful characters that made their homes in the woods. Musicologist and oyster farmer Jeff McKeen of Montville, Maine, has collected songs all over the state, and he remembers a firsthand account of a song used by lumberjacks as they created vast rafts of logs, marching around a capstan to the rhythm of a song. Though the north woods songs are ballads and not in the classic call-and-response format of many worksongs, many are highly rhythmic and would "work" well for swinging the ax. They certainly work well for me in stacking my woodpile.

We don't have the clearest record of Maine farmers singing while actually working, but the archives give us tantalizing glimpses of farms filled with music. Many of the old Irish and English farming songs survived here in slightly modified form and were definitely sung in kitchens and parlors. And the trope of the farmer singing behind the plough is not unfamiliar, as immortalized by old-time Maine poet Holman F. Day in his "Song of the Harrow and the Plow."

The Grange, a proud institution in Maine, has a large collection of songs that almost certainly followed the politically active and community-minded farmer from the hall at night into the field the next morning. These halls were also the site of a staggering number of regular community dances, and many farmers padded their earnings by playing the fiddle and singing the figures for the dancers, who were often their farm customers, colleagues, and coworkers. This music knit together the social network that was the basis of profitable small farms across the state.

Perhaps the best-known Maine farmer's worksong is "'Tis the Gift to Be Simple." It was written in Alfred, Maine, by Shaker elder Joseph Brackett and is often listed as a worksong-hymn. Shaker communities have been a part of Maine culture since the 1700s, and I have heard that the shakers in next-door Canterbury, New Hampshire were known to sing while scything, and at times had a hundred people scything at a time. "'Tis the Gift to Be Simple" is perfectly timed to match the rhythm of scything, and though I have not done the research, I expect we will find evidence of the many fantastic Shaker songs being used, like their furniture, as practical and spiritual tools of startling efficiency.

Although I am not a Shaker, I find the simple act of belting out a song in the field to be transcendent in its own way. Your senses fire on. Bird calls, which are ordinarily just wallpaper, suddenly stand out and take a place in the song. The entire sonic landscape feels ordered—a truck's whoosh becomes a snare drum and an airplane is a drone that you sing against. You are suddenly hearing things clearly, almost for the first time.

And it carries over to the other senses. The soil smells stronger and richer. With each breath you can sense the plants, breathing and

R. de Salis

pollinating and decaying around you. There is an energy in the air. Red pops out against green, purple aligns with yellow. It feels significant.

And at the same time it feels like it just makes sense. These songs, rooted in this place, match the Mainer's classic sense of practicality. Doing good work and having fun doing it, that's a credo, if ever there was one, for Maine farmers, past and present.

To hear song recordings and to learn more about Maine worksongs, visit www.worksongs.org.

IN THIS TIME OF GREAT FORGETTING

RYAN POWER

In this time of Great Forgetting, we choose to remember. To recall our embeddedness in and kinship to the natural world. To remember our deep need of one another, and rediscover what writer Francis Weller calls the "commons of the soul."

Without a cultural tradition to lean on, we turn directly to the elements. We do not simply to go look; we engage, we question, we work—we farm with them.

And on our farms we soon find that it is impossible to forget. The weather, the soil, the wild animals (read: gophers), the plants; they all constantly remind us. We begin to see ourselves, the people, as just part of the symphony—a conductor, perhaps, or a composer, but we alone do not make the music.

The birdsong and the wind and the dirt are common and easy to find. Yet as we engage these we discover something that today is rare.

Farming puts us directly into relation with the living earth. When we have turned our lives over to the seasons and to our farms, we begin to learn and change and understand things that once were just philosophy. We take the cycles of life, death, and rebirth into our daily work. We blend with the land; when we look at any part of the farm, we can see our choices there. We learn to see the seed in the plant. We begin to inhabit the commonality of the earth, within which there is a richness that cannot be bought or sold.

As we grow and relax into that commonness, we find a wilderness within it. And there, we can feel what we share with all of life.

ABOUT THE GREENHORNS

The Greenhorns is a grassroots nonprofit organization made up of young farmers and a diversity of collaborators. Our mission is to recruit, promote, and support the new generation of young farmers. We do this by producing avant-garde programming, video, audio, web content, publications, events, and art projects that increase the odds for success and enhance the profile and social lives of America's young farmers. In addition to the Almanac, our current endeavors include:

UP UP! FARM FILM FEST

Up Up! Farm Film Fest is a film collection with seventeen hours of documentaries by thirteen independent filmmakers (including the Greenhorns) about the future of farming, featuring young farmers from around the world. Each film explores questions of farmland access, rural livelihoods, and the relationships of people and place.

The project aims to bring young farmers' stories to the ears and eyes of entire communities and into the hands of aspiring farmers. Learn more about how to bring Up Up! Farm Film Fest to your community by visiting www .upupfilmfest.org or by emailing film@thegreenhorns.net.

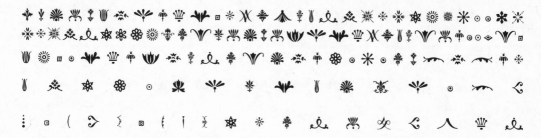

GREENHORNS RADIO

Greenhorns Radio is radio for young farmers, by young farmers. Hosted by acclaimed activist, farmer, and film-maker Severine von Tscharner Fleming, Greenhorns Radio is a weekly phone interview with next-generation farmers and ranchers, surveying the issues critical to their success. We hold no punches. Dig in to our archives or listen live at http://heritageradionetwork.org/series/greenhorns-radio/.

OURLAND TV

We have created this series of films to celebrate and interpret the interventions undertaken by individuals and communities to shift our food and farm economy. Each episodes addresses a major systemic failure of the old food economy, such as toxicity, monoculture, monopoly, inequity, exploitation, drought, vulnerability to climate change, etc. The characteristics of our current system makes it inadequate to our needs; however, our critique counts for nothing if we do not act. Each film introduces us to the people who have identified a point of intervention and have put their shoulder to the task. Check out our episodes at www.ourland.tv/.

SAIL FREIGHT

The Greenhorns' Sail Freight Project was conceived as a trade stunt to provoke conversations about regional food security. More future enactment than historical reenactment, the Greenhorns sees the ocean commons as a place for performance and relationship building. Sail Freight celebrates new routes to market for farm businesses as a way to draw attention to agrarian entrepreneurship and to use the logic of the landscape to orient ourselves in the generations-long project of restoring resilient watersheds, foodsheds, and value chains. Read about our past voyages (Maine and Vermont) at www.thegreenhorns.net.

SUBMISSIONS

In our first issue, Rick Prelinger noted that an almanac characteristically concludes with the promise, "Solution in next year's Almanac." Be part of the solution to all the unanswerable conundrums and impossible paradoxes contained herein by contributing to our next volume!

Please be in touch by March 2017 with inquiries and ideas for our 2019 edition. We're at almanac@thegreenhorns.net. We look forward to hearing from you.

THE NATURE INSTITUTE WINTER INTENSIVE
February 12–17, 2017
The Nature Institute, 20 May Hill Road, Ghent, NY

DEVELOPING A QUALITATIVE UNDERSTANDING OF NATURE:
ANIMALS, HUMANITY, AND EVOLUTION
A course for farmers, apprentices, and educators with
Craig Holdrege, Henrike Holdrege, and Bruno Follador

In light of deep concerns about human, animal, plant, and ecological health—which are all intertwined—the question arises: how can we learn to perceive and discern quality? Science has given us marvelous ways to study the quantitative aspects of nature; we can know all the ingredients of food and what is clearly harmful or a basic prerequisite for survival. But we have lost an intimate qualitative relation to things.

Goethean phenomenology is a means to perceive and understand the qualitative and dynamic aspects of nature. It is an approach that is always grounded in concrete human experience. It strives, through practice, to intensify experience so that we can gain clearer and deeper insights. Such insights can help us to interact with nature in more responsible ways.

For more information: www.natureinstitute.org/educ/winter/index.htm
Contact information: info@natureinstitute.org or (518) 672-0116

357

IMAGE CREDITS

This is a noncommercial, nonprofit publication. All works cited from the Metropolitan Museum of Art are designated Open Access for Scholarly Content (OASC) works. The old newspaper clippings were provided by Tom Giessel. All efforts have been made to provide correct sourcing for materials.

13 Public domain image.

16 Public domain image.

17 Public domain image.

19 Public domain image.

22 *Field Guide to the Acequias of the Middle Rio Grande*, Emily Vogler and Jesse Vogler.

23 Public domain image.

27 Paul S. Taylor (Paul Schuster), 1895–1984. Photograph by Bob Fitch, http://purl.stanford.edu/fq035vf3979.

29 Public domain image.

30 Public domain image.

31 Public domain image.

34 Public domain image.

35 Public domain image.

36 Public domain image.

38 Public domain image.

39 Public domain image.

40 Public domain image.

41 Public domain image.

42 National Oceanic and Atmospheric Association, 2016, www.noaa.gov.

43 National Oceanic and Atmospheric Association, 2016, www.noaa.gov.

49 Stereo card with photo of the moon published by T. W. Ingersoll (1897). Photograph by Prof. Rutherford.

57 Original work by Rachel Alexandrou.

58 Jacob van Ruisdael.

63 Public domain image.

64 Public domain image.

66 "Farm Preserve Note," original work by Martha Shaw; "Deli Dollar," original work by Martha Shaw.

67 Personal cheque, 1955, Barclay's bank; NASA satellite image of the Amazon River mouth, Wikipedia.

69 Sharon Stewart, *Acequia Series*.

71 Rodolph de Salis, "Former Domestic Tin Bath: Light and Portable, Now Useful in the Garden or Field," 2016.

72 Water filter, Wikipedia, 2016.

73 Ats Kurvet, Oculus rift development, 2014, Wikimedia Commons.

75 Original work by Ron Gautreau.

77 Office of the State Engineer, Interstate Stream Commission, Bohannan Huston, January 2014.

78 Original work by Allyson Siwik.

81 Original work by Allyson Siwik.

84 John Rocque, 1761 map of parts of Enborne Wash, Newbury Wash (Wash Common), East Enborne and Sandleford, Berkshire, UK.

87 Original work by Nik Bertulis.

90 Original work by Sarah Gittins.

92 Rodolph de Salis, "Sir Patrick Leigh Fermor's Pebbles," Kalamitsi, Mani, Greece, 2009.

94 Helga von Cramm, "Weisshorn," chromolithograph, 1870s.

95 Helga von Cramm, "Jungfrau," chromolithograph, 1870s.

96 Original work by John Snider.

98 E. O. (Elias Olcott) Beaman, 1837–1876, photographer, James Fennemore, 1849–1941, photographer, John K. Hillers, 1843–1925, photographer, Cataract Canyon, Colorado River. Clem Powell reading while sitting on a boulder in the river. War Department, Office of the Chief of Engineers, Powell Survey. (1869–circa 1874), 1871–1878.

99 Public domain image.

101 Original work by Rachel Alexandrou.

102 Lt. John Warwick Brooke, Shell-carrying pack mules during Battle of Pilckem Ridge, third battle of Ypres, World War I, August 1, 1917.

105 The Metropolitan Museum of Art, Utagawa Hiroshige (Japanese, Tokyo [Edo], 1797–1858, Tokyo [Edo]), "Swallow and Wisteria," www.metmuseum.org.

106 The Metropolitan Museum of Art, Totoya Hokkei (Japanese, 1780–1850), "Young Pine Tree and Jeweled Broom," www.metmuseum.org.

109 The Metropolitan Museum of Art, Katsushika Hokusai (Japanese, Tokyo [Edo], 1760–1849, Tokyo [Edo]), "Winter Landscape," www.metmuseum.org.

111 Original work by Jason Detzel.

113 Frederick Walker, "The Vagrants," ARA, 1868.

IMAGE CREDITS

114 Original work by Ben Short.

118 Wintery field showing a headland, Wikipedia.

120 Original work by Laine Shipley.

122 Original work by Rachel Weaver.

125 Original work by Rachel Alexandrou.

126 Original work by Queen Mob Collective.

130 Rodolph de Salis, "Former Railway," Hampshire, UK, 2016.

132 Public domain image.

133 Public domain image.

135 John Craxton, landscape, lithograph, from Geoffrey Grigson, ed., *Visionary Poems and Passages or The Poet's Eye* (London: Frederick Muller, 1944).

139 www.thehopproject.co.uk.

141 Public domain image.

142 Original work by SPURSE.

144 Public domain image.

145 Original work by Rachel Alexandrou.

146 Rodolph de Salis, "No Public, with *Urtica dioica* (Stinging Nettle)," 2016.

148 Original work by Yoni Gantcher.

151 Public domain image.

152 Rama, "Pole Arms, Swiss," via Wikipedia.

154 Paulus Hector Mair and Jörg Breu the Younger (circa 1510–1547), A flail, as seen in the *fechtbuch* compendium of Paulus Hector Mair (1517–1579).

155 The Metropolitan Museum of Art, "Helmets, Bascinet, War Hat, Comb Morion," www.metmuseum.org.

156 The Metropolitan Museum of Art, "The Tree of Life," first half 17th century, www.metmuseum.org.

159 Original work by Bonnie Rose Weaver.

161 Original work by Bonnie Rose Weaver.

163 Original work by Rachel Alexandrou.

164 Rodolph de Salis, "A Decorative Folly Lodge," Highclere Castle, Hampshire, UK, 2016.

168 Pieter Brueghel the Elder, "Tower of Babel," 1563.

173 Maseltov, "Massive Rheinischer Winterrambur Apple Tree," Burgwald-Ernsthausen, Germany, via Wikipedia.

174 Public domain image.

175 Public domain image.

176 Rodolph de Salis, "Apples," Hampshire, UK.

180 The Metropolitan Museum of Art, Odilon Redon (French, Bordeaux, 1840–1916, Paris), "The Buddha," www.metmuseum.org.

183 Original work by Nick Hayes, www.foghornhayes.com; instagram: nickhayesillustration.

184 Jacob van Ruisdael (1628/29–1682), "Windmill."

186 Public domain image.

187 Public domain image.

188 Public domain image.

189 Public domain image.

190 Public domain image.

191 Public domain image.

192 Public domain image.

193 Public domain image.

194 Original work by Martha Shaw.

195 Original work by Rachel Alexandrou.

196 The Metropolitan Museum of Art, Eadweard Muybridge (American, born Britain, 1830–1904), "Animal Locomotion. An Electro-Photographic Investigation … of Animal Movements. Commenced 1872—Completed 1885. Volume VII, Men and Woman (Draped) Miscellaneous Subjects," www.metmuseum.org.

199 Public domain image.

201 Silver ducat from the time of the last Doge of Venice, Ludovicus Manin. Scan by Rodolph de Salis.

204 Anonymous, "Labours of the Month (July)," 15th century, stained glass.

206 Public domain image.

207 Anonymous, "Anglo-Saxon Ploughing."

210 William M. Van der Weyde (American, 1870–1929), "Timberwolves," circa 1900, via George Eastman House, Flickr Commons Archive.

211 Public domain image.

213 Lyn Archer, *Three Strippers at Devils Point Bar, Portland, OR*, 2006.

217 Public domain image.

218 Public domain image.

219 Original work by Rachel Alexandrou.

220 The Metropolitan Museum of Art, Albrecht Dürer (German, Nuremberg, 1471–1528, Nuremberg), "Underweysung der messung mit dem zirckel un richt scheyt," www.metmuseum.org.

224 Frederick W. W. Howell, Reykjavik. "National

IMAGE CREDITS ✳ ✳ ⬯ ⸮ ▫ ✳ ❦ ⚭ ✱ ⊙ ✳ ❀ ◦ ❁ ⚘ ✱ ✳ ❋ ◦ ✿

Celebration," 1898, collodion print, via Cornell University Library, Flickr Commons Archive.

226 *The Kansas Union Farmer* newspaper clip was recaptured by Tom Giessel, Honorary Historian of National Farmers Union.

228 Hans Lützelburger, "Sixteenth-Century Fighting in Bavaria," by Hans Lützelburger.

230 Public domain image.

231 "Fly to the Moon," via Flickr Creative Commons, user fdecomite, 2007.

232 Rodolph de Salis, "Apples."

234 Public domain image.

235 Taber Photographic Parlors, circa 1882–1888, "Big Trees, Felton, Santa Cruz Co.," DeGolyer Library, via SMU Central University Libraries, Flickr Commons Archive.

236 René Descartes, "Die drei Spären der Erde," *Principia Philosophiae* IV:6, 1644.

239 The Metropolitan Museum of Art, John Dillwyn Llewelyn (British, Swansea, Wales, 1810–1882, Swansea, Wales), "Thereza Dillwyn Llewelyn with Her Microscope," www.metmuseum.org.

240 Pieter Bruegel the Elder (circa 1525–1569), "The Corn Harvest (August)," 1565.

244 Rodolph de Salis, "Ancient Boundary Wood Bank and Modern Fence between Greenham Common and Peckmore Coppice," Berkshire, UK, 2016.

246 Public domain image.

247 Original work by Rachel Alexandrou.

248 Public domain image.

253 Frederick W. W. Howell, "Creamery (Rjómabú) at Seljaland (Eyjafjoll)," 1911, via Cornell University Library, Flickr Commons Archive.

255 William H. Martin (1865–1940), "Feeding Time," 1909, via George Eastman House, Flickr Commons Archive.

257 Public domain image.

260 The Metropolitan Museum of Art, Katsushika Hokusai (Japanese, Tokyo [Edo], 1760–1849, Tokyo [Edo]), "Album of Sketches by Katsushika Hokusai and His Disciples," www.metmuseum.org.

263 Hélvetius, Jean-Claude-Adrien, "De l'esprit … A Paris, chez Durand, libraire, Kolofon, De l'imprimerie de Moreau, imprimeur de la Reine," 1758.

264 The Metropolitan Museum of Art, Adriaen van Ostade (Dutch, Haarlem, 1610–1685, Haarlem), "The Pissing Man," 1610–1685, www.metmuseum.org.

266 Rodolph de Salis, "Corn," Hampshire, UK.

269 Public domain image.

271 Public domain image.

274 Original work by Chong Jones.

275 Image courtesy of Charlotte Sullivan.

277 Original work by Rachel Alexandrou.

278 Giovanni Segantini (1858–1899), "Mezzogiorno sulle Alpi (Midday in the Alps)," 1891.

280 Apple Creek Farm, http://lwrncbrn.pixieset.com/applecreekfarm/lores/.

281 Frederic, Lord Leighton (1830–1896), *Flaming June*, 1895.

283 Original work by Daniel Tucker.

285 Original work by Daniel Tucker.

286 "Spherical Harmonic of the Third Order," from James Clark Maxwell, *A Treatise on Electricity and Magnetism* (Oxford, UK: Clarendon Press, 1973).

289 Public domain image.

291 Public domain image.

294 Charles Darwin, 1837.

295 Original work by Rachel Alexandrou.

296 "Macgregor, Owned by James R. Dempster," Ladyton, 1890, via National Galleries of Scotland, Flickr Commons Archive.

298 Louis Ducos Du Hauron (1837–1920), *Still Life with Rooster*, circa 1869–1879, via George Eastman House, Flickr Commons Archive.

301 The Metropolitan Museum of Art, Utagawa Toyoharu (Japanese, 1735–1814), Cock, Hen, and Chicken, www.metmuseum.org.

303 The Metropolitan Museum of Art, Attributed to Julius Wiesner (Austrian, 1838–1916), "Frustules of Diatoms," www.metmuseum.org.

305 Public domain image.

308 Public domain image.

309 Public domain image.

312 The Metropolitan Museum of Art, Samuel H. Gottscho (American, 1875–1971), "Trylon and Perisphere, New York World's Fair," www.metmuseum.org.

314 Patent No. 5,255,452, "Michael Jackson's Anti-Gravity Illusion Shoes," Selected Patent Case Files

IMAGE CREDITS

Record Group 241, Records of the Trademark and Patent Office, 10/26/199, via The U.S. National Archives, Flickr Commons Archive.

315 Patent No. 5,255,452, "Michael Jackson's Anti-Gravity Illusion Shoes," Selected Patent Case Files Record Group 241, Records of the Trademark and Patent Office, 10/26/199, via The U.S. National Archives, Flickr Commons Archive.

317 Rodolph de Salis, "Off-Grid: Asbestos-Sheet-Coated Barn Once Used to Generate Electricity," Hampshire, UK, 2006.

318 Public domain image.

319 Public domain image.

323 Original work by Rachel Alexandrou.

324 Rodolph de Salis, "Ancient Yew," Ecchinswell, Hampshire, UK, 2016.

327 Rodolph de Salis, "Ash Tree," North Yorkshire, UK, 2015.

328 Original work by Gisela Weimann (Benutzername Charlotte Huhn).

330 Department of the Interior. Patent Office. (1849–1925), "Drawing of Rocking Chair and Fan," 08/07/1847–08/07/1847, via The U.S. National Archives, Flickr Commons Archive.

332 Public domain image.

333 Public domain image.

336 The Metropolitan Museum of Art, Unidentified Artist, Japanese, "Mynah Birds," www.metmuseum.org.

340 The Metropolitan Museum of Art, Katsushika Hokusai (Japanese, Tokyo, [Edo], 1760–1849, Tokyo [Edo]), "Grasshopper and Iris," www.metmuseum.org.

341 Public domain image.

345 The Metropolitan Museum of Art, Vincent van Gogh (Dutch, Zundert, 1853–1890, Auvers-sur-Oise), "Wheat Field," www.metmuseum.org.

347 Public domain image.

348 Public domain image.

349 Public domain image.

350 Public domain image.

360 Public domain image.

NOTES

NOTES

NOTES

NOTES

NOTES

NOTES

Comptoir
ALMANACH
voor het Schrickel-Iaer onses
Heeren duysent ses hondert
ende sesthien,

Nae den Nieuwen ende Ouden Stijl.

Ghecalculeert op den Meridiaen der
vermaerder Koop-stadt Amsterdam.
Door Lucas Iansz. Sinck.

Op desen Almanach is gestelt hoe die ou-
de Roomsche Keysers plachten vergodet te worden,
het triumpheren der Roomsche Overwinners/
ende eenighe gheneuchlijcke kluchten.

t'AMSTERDAM, By Baer Iansz.
Voor Hendric Laurentsz int Schrijf-boec.